Electromagnetic Wave Propagation for Industry and Biomedical Applications

Edited by Lulu Wang

Published in London, United Kingdom

IntechOpen

Supporting open minds since 2005

Electromagnetic Wave Propagation for Industry and Biomedical Applications
http://dx.doi.org/10.5772/intechopen.87687
Edited by Lulu Wang

Contributors
Fred Lacy, Rafael Zamorano Ulloa, Abdelhak Hafdallah, Mouna Abdelli, Shiva Hayati Raad, Zahra Atlasbaf, Sukhmander Singh, Ashish Tyagi, Bhavna Vidhani, Ling Zhang, Er-Ping Li, Yuru Feng, Jun Fan, Renata Rampim, Ivan de Pieri Baladei, Narges Kasiri, Mohammed Kanjaa, Otman El Mrabet, Mohsine Khalladi

First published in London, United Kingdom, 2022 by IntechOpen
IntechOpen is the global imprint of INTECHOPEN LIMITED, registered in England and Wales, registration number: 11086078, 5 Princes Gate Court, London, SW7 2QJ, United Kingdom
Printed in Croatia

British Library Cataloguing-in-Publication Data
A catalogue record for this book is available from the British Library

Additional hard and PDF copies can be obtained from orders@intechopen.com

Electromagnetic Wave Propagation for Industry and Biomedical Applications
Edited by Lulu Wang
p. cm.
Print ISBN 978-1-83968-581-1
Online ISBN 978-1-83968-582-8
eBook (PDF) ISBN 978-1-83968-583-5

For EU product safety concerns:
IN TECH d.o.o., Prolaz Marije Krucifikse Kozulić 3, 51000 Rijeka, Croatia, info@intechopen.com or visit our website at intechopen.com.

Meet the editor

Lulu Wang is a Distinguished Professor of Biomedical Engineering, Shenzhen Technology University, China. Dr. Wang is a member of the American Society of Mechanical Engineers (ASME), Institute of Electrical and Electronics Engineers (IEEE), American Association for the Advancement of Science (AAAS), Physiological Society of New Zealand (PSNZ), and Institution of Professional Engineers New Zealand (IPENZ). Her research interests include medical devices, electromagnetic sensing and imaging, and computational mechanics. Over the past five years, Dr. Wang has authored more than seventy peer-reviewed publications, two ASME books, seven book chapters, and ten issued patents. She is an active reviewer of numerous journals, books, and conferences. She has edited four books and three special issues of international journals. She has received multiple national and international awards from various professional societies and organizations. She is an active organizer of several international conferences, including the ASME International Mechanical Engineering Congress & Exposition and the International Conference on Computational & Experimental Engineering Sciences.

Contents

Preface

This book presents recent achievements in electromagnetic wave propagation for industry and biomedical applications by covering design, modeling, and experimentation aspects. It discusses novel approaches to electromagnetic imaging, electromagnetic devices, wireless implants, and in vitro and in vivo testing.

This book consists of three sections. The first section provides electromagnetic wave propagation fundamental field equations, the second section demonstrates the development of electromagnetic wave radiation and scattering, and the third section focuses on the development of electromagnetic wave propagation, radiation and scattering applications.

Section 1, "Electromagnetic Wave Propagation Fundamental Field Equations," consists of three chapters. Dr. Hayati et al. presents Green's function derivation for multilayered planar, cylindrical, and spherical structures with impedance boundary condition. Dr. Kanjaa et al. provide a TLM formulation based on fractional derivatives for dispersive Cole-Cole media. Dr. Hafdallah et al. present averaged no-regret control for an electromagnetic wave equation depending upon a parameter with incomplete initial conditions.

Section 2, "Electromagnetic Wave Radiation and Scattering," consists of two chapters. Dr. Lacy provides a concise introduction of identifying superconducting materials through electromagnetic properties, and Dr. Singh et al. demonstrate the physics of absorption and generation of electromagnetic radiation.

Section 3, "Electromagnetic Wave Propagation, Radiation and Scattering Applications," consists of four chapters. Dr. Zamorano et al. present electromagnetic features of microwave heating, Dr. Zhang et al. present high-frequency electromagnetic interference diagnostics, and Dr. Rampim et al. describe UHF RFID development in a harsh metallic environment. Dr. Kasiri presents RFID and its applications in retail.

I would like to express my sincere thanks to all the contributing authors.

Lulu Wang
Shenzhen Technology University,
Shenzhen, China

Section 1

Electromagnetic Wave Propagation Fundamental Field Equations

Chapter 1

Dyadic Green's Function for Multilayered Planar, Cylindrical, and Spherical Structures with Impedance Boundary Condition

Shiva Hayati Raad and Zahra Atlasbaf

Abstract

The integral equation (IE) method is one of the efficient approaches for solving electromagnetic problems, where dyadic Green's function (DGF) plays an important role as the Kernel of the integrals. In general, a layered medium with planar, cylindrical, or spherical geometry can be used to model different biomedical media such as human skin, body, or head. Therefore, in this chapter, different approaches for the derivation of Green's function for these structures will be introduced. Due to the recent great interest in two-dimensional (2D) materials, the chapter will also discuss the generalization of the technique to the same structures with interfaces made of isotropic and anisotropic surface impedances. To this end, general formulas for the dyadic Green's function of the aforementioned structures are extracted based on the scattering superposition method by considering field and source points in the arbitrary locations. Apparently, by setting the surface conductivity of the interfaces equal to zero, the formulations will turn into the associated problem with dielectric boundaries. This section will also aid in the design of various biomedical devices such as sensors, cloaks, and spectrometers, with improved functionality. Finally, the Purcell factor of a dipole emitter in the presence of the layered structures will be discussed as another biomedical application of the formulation.

Keywords: biomedical, dyadic Green's function, integral equation, 2D materials, sensors, Purcell factor

1. Introduction

Planarly, cylindrically and spherically layered media have been widely used to model the human skin, body, or head. In particular, a rectangular slab is proposed to model and analyze skin temperature distribution [1]. Moreover, in the multilayered skin model, three stacked layers are exploited to simulate the performance of the epidermis, dermis, and sub-cutis parts of the skin [2]. In other research, a planarly layered medium has been proposed as the simplified human body model by considering the impact of skin, fat, and muscle in the electromagnetic performance [3]. Cylindrical-shaped equivalent phantom of the skin is alternatively used to characterize the interactions between an antenna and the human body [4]. For a more precise investigation, multilayered cylinders are proposed to model a biological system with different tissues [5]. Considering spherical geometrics, the interactions of a three-layered spherical human head model with a finite-length dipole and a cellular phone helical antenna are investigated using Green's function [6, 7]. The research is extended to the six-layered model, including skin, fat, bone, dura, CSF, and brain layers of the head [8]. This model has also been used to explore the impact of the shapes and positions of coils in the MRI system [9]. For the analysis of the aforementioned structures, integral equation (IE) based methods have been widely used. Dyadic Green's function, which is a powerful method to calculate the electromagnetic response for different excitation sources, plays an essential role in the IE method [10–12].

Different mathematical approaches can be utilized to calculate the dyadic Green's function. The dyadic Green's function in a homogeneous environment can be expressed in terms of vector functions $\boldsymbol{M}$, $\boldsymbol{N}$, and $\boldsymbol{L}$. The Green's function is singular when the field and observation points coincide. In this case, Green's function can be written as the main part plus the portion proportional to the impulse function. This type of decomposition of the Green's function is not unique and depends on the shape of the volume that separates the environment [13]. In general, to obtain the dyadic Green's function in a layered medium, the Green's function of the homogeneous environment can be used. Then, the effect of inhomogeneity can be considered by adding reflected and transmitted waves to the dyadic Green's function [14]. In a similar approach, called the scattering superposition method, the scattering Green's function in each layer is expressed in terms of vector wave functions with the unknown coefficients that are obtained by applying the boundary conditions [15]. Moreover, to calculate the Green's function of a medium by impedance method, instead of each dielectric layer a transmission line, and instead of each metallic layer a current source can be considered in the equivalent model in the rectangular, cylindrical, and spherical coordinates [15–17].

Considering medical diagnostics and treatment, planarly, cylindrically and spherically layered media are engineered mainly with plasmonic materials. In this regard, the interaction of an environment with a nanometer-scale dipole emitter is of interest in different biomedical fields. For example, the optical activity of the proteins can be investigated using Green's tensor approach. Furthermore, a single excited molecule in the vicinity of a metallic structure can be potentially used in the sensors because of behaving as a resonant filter. Moreover, the point source is an appropriate model for the concentrated light sources which are used in medical applications [18]. Following these trends, different plasmonic structures have been presented theoretically and realized in the real environment. For example, the effect of the size of the gold nanoparticles on the decay rate and the energy transfer of dipole emitters is investigated using Green's tensor formulation and compared with the results obtained from the fabrication [19]. The same analysis is carried out for a metallic cylinder coated with a dielectric layer and the resulted integrals are solved numerically [20]. Also, Fermat's golden rule is used to connect the imaginary part of the Green's function to the radiation impedance of the dipole antenna adjacent to a medium, and the Purcell factor is extracted for the planar meta-surface, which is then experimentally characterized in the microwave frequency band [21].

Recently, graphene's plasmons are proposed as the low loss and reconfigurable alternative to the plasmons of the noble metals. There are two approaches to use graphene in analytical and numerical methods. In the first one, the graphene boundary is modeled with a 2D surface, characterized by its surface conductivity, whereas in the second one, the graphene layer is replaced by a very thin dielectric [22]. Although the latter can be analyzed using the available formulas previously presented for the multilayered dielectric structures in the literature [13, 23], graphene analysis using surface conductivity model has many advantages. First, in the dielectric model, each graphene interface adds an extra layer to the structure. Therefore, when the number of graphene layers is large, the problem becomes very complicated. Secondly, in the dielectric model, it is necessary to compute the special functions of the cylindrical and spherical coordinates with complex arguments, which requires the implementation of specific algorithms for their effective calculation [24], while in the impedance boundary condition method, the surface conductivity of graphene appears as a coefficient for special functions. Third, in the dielectric model, the thickness of the graphene layer is considered to be about 0.335 nm, which is often very small compared to other geometrical parameters and makes the convergence of the analytical functions slow [25]. Also, if the goal is using the dielectric model in numerical methods, it is necessary to use a dense mesh for the equivalent dielectric of the graphene that is not optimal in terms of time and memory [26].

Applying the graphene surface conductivity model for the derivation of the Green's function has been considered in recent years. For example, for the graphene sheet under electric bias, the dyadic Green's function is derived by the Hertzian potential and plane-wave expansion methods and the corresponding integrals are solved with the saddle point method [27, 28]. Also, the method is expanded for the analysis of graphene with tensor surface conductivity which can be used to analyze graphene with magnetic bias or spatial dispersion. Romberg's integration procedure is proposed for the numerical solution of the resulting integrals [29]. In another research, the analysis of the electric dipole in the proximity of the parallel plate waveguide with graphene walls is studied by extracting the Green's function and calculating the spontaneous emission. It is found that symmetric and asymmetric plasmonic modes lead to a sharp increase in this parameter [30]. As another instance, a point source is taken into account in the vicinity of the infinite cylinder with graphene cover and it is observed that by changing the distance of the source from the cylinder as well as changing the chemical potential of the graphene layer, the Fano resonances can be controlled [31]. In this chapter, the dyadic Green's function of various planar, cylindrical, and spherical geometries with impedance boundary conditions will be calculated using the scattering superposition method. Specifically, we have focused on the graphene material due to its wide range of applications. Apparently, another 2D material can be considered by replacing the graphene surface conductivity with the surface conductivity of the desired material. The presented formulas can be potentially sued to design various biomedical devices. Moreover, by approaching the surface impedance to zero, these structures can be potentially used to investigate the interaction of the human body with different electromagnetic sources. For the Green's function calculation of complex media in different coordinates using vector wave functions, the reader is referred to [32–35].

2. Surface conductivity of graphene material under different conditions

Graphene is a two-dimensional material made of carbon atoms and can be considered in the solution of Maxwell's equations using surface conductivity boundary condition [23]. Depending on the geometrical and optical conditions, graphene surface conductivity can be isotropic or anisotropic. The purpose of this section is to provide an overview of the graphene surface conductivity under different conditions (electric bias, magnetic bias, and spatial dispersion) and for different geometries (continuous or patterned sheets).

It should be noted that graphene material is mainly synthesized through four methods, including 1) mechanical exfoliation of highly ordered pyrolytic graphite (HOPG), 2) the epitaxial growth of graphene on silicon carbide (SiC), 3) the reduction of graphene oxide, and 4) chemical vapor deposition (CVD) technique. The comparison of these methods in terms of quality and the fabricated area is provided in **Table 1** [36].

Method	Quality	Area
Mechanical exfoliation of highly ordered pyrolytic graphite (HOPG)	Very high	Small
Epitaxial growth of graphene on silicon carbide (SiC)	Medium	Large (3–4 inches wafers)
Reduction of graphene oxide (rGO)	Medium	Large
Chemical vapor deposition (CVD)	High	Very large (30 inches)

Table 1.
Comparison of the quality and area of synthesized graphene using different techniques [36].

In this regard, a monolayer graphene film with metallic electrodes is transferred to a high impedance surface (HIS) which is realized by the printed circuit board (PCB) technology at microwave frequencies. The measurement is conducted in a small microwave chamber [37]. In another research in the same spectrum, an infrared laser is used to etch the CVD grown graphene sheet with predefined periodicity and later investigate the absorption of the designed structure inside a rectangular waveguide [38]. The electron beam lithography is another approach used for pattering the graphene sheet for enhanced light matter interaction [39]. Moreover, a transparent graphene millimeter wave absorber constructed by multiple transfer-etch processing is characterized by reflectometery technique at 140 GHz [40]. Also, CVD-grown graphene is used to enhance the sensitivity of the surface-enhanced Raman spectroscopy (SERS)-based chemical sensor. The measurement is done using a Raman spectrometer at the laser wavelength of 785 nm (red) [41]. In another sensor chip, DNA is hybridized to the graphene-based substrate under UV light with the wavelength of 260 nm. In this sensor, the atomic force microscopy (AFM) is used to ensure the continuity and uniformity of the synthesized graphene, and Raman characterization is used to investigate its quality and the number of layers [42].

2.1 Graphene material under electric bias

When a graphene sheet is under electric bias, its surface conductivity is isotropic and can be approximately calculated by using Kubo's formulas as [43]:

$$\sigma_{\text{intra}} = \frac{2ie^2 k_B T}{\hbar^2 \pi(\omega + i/\tau)} \ln\left[2\cosh\left(\frac{\mu_c}{2k_B T}\right)\right] \tag{1}$$

$$\sigma_{\text{inter}} = \frac{e^2}{4\hbar} \times \left[\frac{1}{2} + \frac{1}{\pi}\arctan\left(\frac{\hbar\omega - 2\mu_c}{2k_B T}\right) - \frac{i}{2\pi}\ln\left(\frac{(\hbar\omega + 2\mu_c)^2}{(\hbar\omega - 2\mu_c)^2 + (2k_B T)^2}\right)\right] \tag{2}$$

In the above equations, T is the temperature, μ_c is the chemical potential of graphene, $\hbar$ is the reduced Planck's constant, K_B is the Boltzmann's constant, and ω is the angular frequency. At low-THz frequencies, the inter-band contribution of the surface conductivity can be neglected. Also, the graphene layer can be modeled as a dielectric with very low thickness δ and equivalent dielectric constant $\varepsilon = 1 - i\frac{\sigma}{\omega\varepsilon_0\delta}$ [22]. Note that in the above equations, graphene is assumed to be in the linear region, otherwise, other terms proportional to $1/\omega^3$ and $1/\omega^4$ should be added respectively for the frequency range of $\hbar\omega < 2\mu_c$ and $\hbar\omega \geq 2\mu_c$ [44].

To increase the light-matter interaction of graphene material, the nano-pattering method has been proposed [45]. The surface conductivity of the periodic graphene

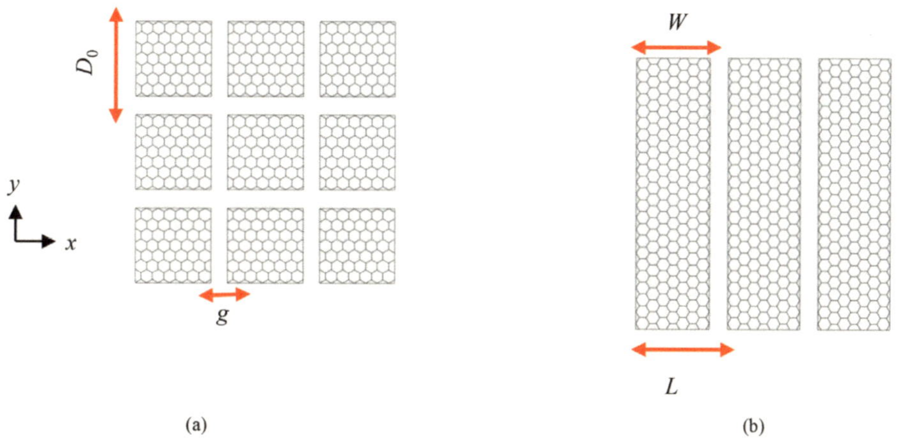

Figure 1.
(a) Graphene square patches with the periodicity of D_o *and the air gap distance of* g *and (b) densely packed graphene strips with the width* W *and the periodicity* L *[46].*

elements under the electric bias is also isotropic. For the square patches shown in **Figure 1(a)**, by considering the periodicity of D_0 and the air gap distance of g, closed-form surface impedance (inverse of the surface conductivity) is as [47]:

$$Z_s = \frac{D_0}{\sigma_s(D_0 - g)} + i\frac{\pi}{2\omega\varepsilon_0\left(\frac{\varepsilon_2+1}{2}\right)D_0 \ln\left[\csc\left(\frac{\pi g}{2D_0}\right)\right]} \tag{3}$$

The effective surface conductivity of periodic graphene elements with arbitrary pattern can be measured or extracted using the parameter retrieval method through full-wave simulation [48]. It should be noted that for computing the electric field required for each considered chemical potential, approximate equations can be derived as [49]:

$$\mu_c(eV) = \begin{cases} \lambda_1 E_0^{\lambda_2}, E_0(V/nm) \geq 0 \\ -\lambda_1 E_0^{\lambda_2}, E_0(V/nm) < 0 \end{cases} \tag{4}$$

where, $\lambda_1 = 0.3677$ and $\lambda_2 = 0.5010$. For the chemical potentials in the range of [−1,1] eV, the required bias fields are in the order of several volts per nanometer, which can be implemented practically [29].

The surface conductivity of densely packed graphene strips, as illustrated in **Figure 1(b)**, is anisotropic and can be approximated in the form of a diagonal tensor using the effective medium formulation as [50]:

$$\begin{aligned} \sigma_{xx} &= \frac{W\sigma\sigma_C}{L\sigma_C + W\sigma} \\ \sigma_{yy} &= \sigma\frac{W}{L} \\ \sigma_{xy} &= \sigma_{yx} = 0 \end{aligned} \tag{5}$$

In the above equations, W and L are the width and periodicity of the strips, respectively. Also, σ is the surface conductivity of graphene under electric bias, and σ_c is the static conductivity of the surface. Two very important properties of this environment are the existence of near-zero surface conductivity and hyperbolic dispersion region with the potential applications in 2D lens structures and the spontaneous emission enhancement of the dipole emitters [50, 51].

2.2 Graphene sheet with spatial dispersion effects

When the graphene sheet is placed on a substrate with a high dielectric constant, its surface conductivity is a tensor in which the elements depend on the wave propagation constant in the structure. This is called the spatial dispersion effect and the associated formulae for calculating surface conductivity are [29]:

$$\sigma_{xx} = \sigma + \alpha \frac{d^2}{dx^2} + \beta \frac{d^2}{dy^2} \tag{6}$$

$$\sigma_{yy} = \sigma + \beta \frac{d^2}{dx^2} + \alpha \frac{d^2}{dy^2} \tag{7}$$

$$\sigma_{xy} = \sigma_{yx} = 2\beta \frac{d^2}{dxdy} \tag{8}$$

The parameters σ, α, and β are extracted for an unbiased sheet ($\mu_c = 0$) thorough perturbation theory [29]. The resulting equations are valid for the electrically biased sheet ($\mu_c \neq 0$), as well [52].

2.3 Graphene sheet under magnetic bias

When a graphene sheet is under magnetic bias, its surface conductivity is also a tensor. The diagonal elements of this tensor are equal and the off-diagonal elements are opposite in sign defined as [53].

$$\sigma_{xx}(\omega, B_0) = \sigma_0 \frac{1 - i\omega\tau}{(\omega_c\tau)^2 + (1 - i\omega\tau)^2} \tag{9}$$

$$\sigma_{yx}(\omega, B_0) = \sigma_0 \frac{\omega_c\tau}{(\omega_c\tau)^2 + (1 - i\omega\tau)^2} \tag{10}$$

where $\sigma_0 = \frac{2e^2\tau}{\pi\hbar^2} k_B T \ln\left(2\cosh\frac{\mu_c}{2k_B T}\right)$ and $\omega_c = \frac{eB_0 v_F^2}{\mu_c}$. The approximate formulas for the calculation of the surface impedance of the square graphene elements under magnetic bias, shown in **Figure 1(a)**, are as follows [54]:

$$\overline{\overline{Z}}_p = F_G \overline{\overline{Z}}_g + \frac{i}{2\alpha}\sqrt{\frac{\mu_0}{\varepsilon_0 \varepsilon_{eff}}}\begin{bmatrix} 1 & 0 \\ 0 & 1 \end{bmatrix} \tag{11}$$

where $F_G = 0.6[D_0/(D_0 - g)]^3 + 0.4$ and $\alpha = -\frac{k_0 D_0 \sqrt{\varepsilon_{eff}}}{\pi} \ln\left[\sin\left(\frac{\pi g}{2D_0}\right)\right]$. Similar to the electrically biased patterned elements with arbitrary shapes, the parameter retrieval method can be used under the applied magnetic bias [55]. Given the discussion of the above three sections, it is observed that assuming the surface conductivity of graphene as:

$$\overline{\overline{\sigma}} = \begin{bmatrix} \sigma_{xx} & \sigma_{xy} \\ \sigma_{yx} & \sigma_{yy} \end{bmatrix} \tag{12}$$

All items expressed above can be extracted as a special case.

3. Analysis of graphene-based structures using dyadic Green's function

Dyadic Green's functions for the planarly, cylindrically, and spherically layered structures with graphene interfaces will be derived in this section. To this end, the boundary conditions of the continuity and discontinuity of tangential electric and magnetic fields are respectively satisfied regarding the considered surface conductivity model for the graphene.

3.1 Graphene-based planarly layered media

This section aims to obtain dyadic Green's functions for planar structures with graphene boundaries. This problem can be solved either in the rectangular or cylindrical coordinates. In the first sub-section, a graphene sheet with the tensor surface conductivity boundary condition (TSCBC) is considered and its dyadic Green's function is calculated in the rectangular coordinates. In this case, the anisotropy of the surface impedance causes the coupling of the transverse electric (TE) and transverse magnetic (TM) fields. As a result of coupling, the number of unknown coefficients in the expansion of the dyadic Green's function is increased concerning the electrically biased sheet. In the second part of this section, a graphene-dielectric stack with an arbitrary number of layers is investigated considering the electric bias for the graphene sheets. This problem is solved in the cylindrical coordinates to simplify the calculation of the resulted Sommerfeld integrals.

3.1.1 Graphene sheet with the tensor surface conductivity boundary condition

The purpose of this section is to obtain the dyadic Green's function of a graphene sheet with the tensor surface conductivity in the interface of half-spaces, as shown in **Figure 2(a)**. The constitutive parameters of the top and bottom regions are considered as (ε_1, μ_1) and (ε_2, μ_2), respectively. Without losing the generality of the problem, the source is assumed to be in the first environment and the graphene boundary is considered in $z = 0$ interface. Dyadic Green's function of this structure will be calculated using the scattering superposition method. For this purpose, the Green's function in each region of the problem is written in the form of the Green's function in the absence and presence of structure. Thus [23]:

$$\overline{\overline{G}}_e^{(11)} = \overline{\overline{G}}_{e0}\left(\overline{R},\overline{R'}\right) + \overline{\overline{G}}_{es}^{(11)}\left(\overline{R},\overline{R'}\right) \quad z>0 \tag{13}$$

$$\overline{\overline{G}}_e^{(21)} = \overline{\overline{G}}_{es}^{(21)}\left(\overline{R},\overline{R'}\right) \quad z<0 \tag{14}$$

where $\overline{\overline{G}}_{es}^{(11)}$ and $\overline{\overline{G}}_{es}^{(21)}$ are respectively the scattering Green's function in regions 1 and 2 and they are expanded in terms of $\overline{M}$ and $\overline{N}$ vector wave functions.

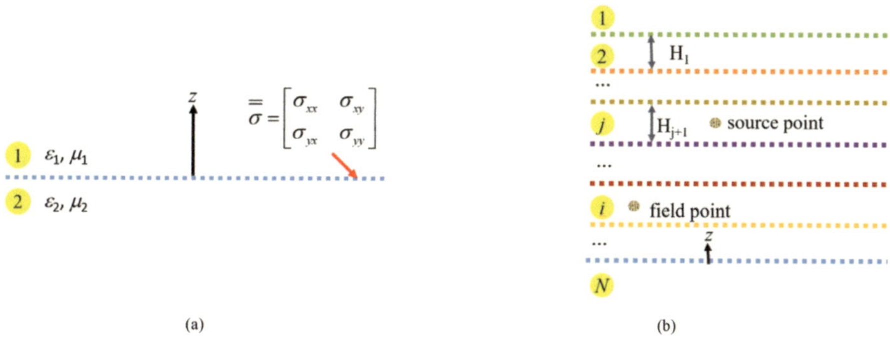

Figure 2.
(a) Graphene sheet with the tensor surface conductivity boundary condition at the interface of half-spaces [46] and (b) graphene-dielectric stack.

Also, $\overline{\overline{\mathbf{G}}}_{e0}$ is the free-space Green's function which can be computed using the $\boldsymbol{G_m}$ method as:

$$\overline{\overline{\mathbf{G}}}_{e0}\left(\overline{\boldsymbol{R}},\overline{\boldsymbol{R}'}\right) = -\frac{1}{(k_1)^2}\hat{\boldsymbol{z}}\hat{\boldsymbol{z}}\,\delta\left(\overline{\boldsymbol{R}},\overline{\boldsymbol{R}'}\right) + \int_{-\infty}^{\infty}\int_{-\infty}^{\infty} dk_x dk_y C^{(1)} \times \left\{\overline{\boldsymbol{M}}(-h_1)\overline{\boldsymbol{M}}'(h_1) + \overline{\mathbf{N}}(-h_1)\overline{\mathbf{N}}'(h_1)\right\} \quad z<z' \tag{15}$$

where $C^{(1)} = \frac{i}{8\pi^2 h_1\left(k_x^2+k_y^2\right)}$. Using the scalar wave function $\psi\left(\overline{k}\right) = \exp\left(ik_x x + ik_y y + ik_z z\right)$, it can be readily found that:

$$\overline{\boldsymbol{M}}\left(\overline{k}\right) = i\left(k_y\hat{\boldsymbol{x}} - k_x\hat{\boldsymbol{y}}\right)\psi\left(\overline{k}\right) \tag{16}$$

$$\overline{\mathbf{N}}\left(\overline{k}\right) = \left(-\frac{k_z}{k_j}\left(k_x\hat{\boldsymbol{x}} + k_y\hat{\boldsymbol{y}}\right) + \frac{k_x^2+k_y^2}{k_j}\hat{\boldsymbol{z}}\right)\psi\left(\overline{k}\right) \tag{17}$$

The parameters k_x, k_y, and k_z are the wavenumbers in x, y, and z directions, respectively, and k_j shows the wavenumber for j = 1, 2. These parameters are not independent and are related to each other via $k_z = \pm\sqrt{k_j^2 - k_x^2 - k_y^2} = \pm\sqrt{k_j^2 - k_\rho^2}$. The vector wave function $\overline{\boldsymbol{M}}$ represents the electric field of TE modes and the vector wave function $\overline{\mathbf{N}}$ shows the electric field of the TM waves. In the structure under consideration, the anisotropy of surface conductivity leads to the coupling of TE and TM fields. Therefore [56]:

$$\overline{\overline{\mathbf{G}}}_{es}^{(11)}\left(\overline{\boldsymbol{R}},\overline{\boldsymbol{R}'}\right) = \int_{-\infty}^{\infty}\int_{-\infty}^{\infty} dk_x dk_y C^{(1)} \times \left[\left(a_1\overline{\boldsymbol{M}}(h_1) + a_1'\overline{\mathbf{N}}(h_1)\right)\overline{\boldsymbol{M}}'(h_1) + \left(b_1\overline{\mathbf{N}}(h_1) + b_1'\overline{\boldsymbol{M}}(h_1)\right)\overline{\mathbf{N}}'(h_1)\right] \tag{18}$$

$$\overline{\overline{\mathbf{G}}}_{es}^{(21)}\left(\overline{\boldsymbol{R}},\overline{\boldsymbol{R}'}\right) = \int_{-\infty}^{\infty}\int_{-\infty}^{\infty} dk_x dk_y C^{(1)} \times \left[\left(a_2\overline{\boldsymbol{M}}(-h_2) + a_2'\overline{\mathbf{N}}(-h_2)\right)\overline{\boldsymbol{M}}'(h_1) + \left(b_2\overline{\mathbf{N}}(-h_2) + b_2'\overline{\boldsymbol{M}}(-h_2)\right)\overline{\mathbf{N}}'(h_1)\right] \tag{19}$$

The unknown coefficients a_1, a_1', b_1, b_1', a_2, a_2', b_2, and b_2' will be obtained by applying the boundary conditions. Using the self and mutual orthogonality of the vector wave functions, the above equations can be divided into two systems of equations, each with four unknown coefficients. The boundary conditions on the electric and magnetic Green's functions respectively state that:

$$\hat{\boldsymbol{z}} \times \left(\overline{\overline{\mathbf{G}}}_e^{(11)} - \overline{\overline{\mathbf{G}}}_e^{(21)}\right) = 0 \tag{20}$$

$$\hat{\boldsymbol{z}} \times \left(\frac{\nabla \times \overline{\overline{\mathbf{G}}}_e^{(11)}}{i\omega\mu_1} - \frac{\nabla \times \overline{\overline{\mathbf{G}}}_e^{(21)}}{i\omega\mu_2}\right) = \overline{\overline{\boldsymbol{\sigma}}}.\overline{\overline{\mathbf{G}}}_e^{(21)} \tag{21}$$

After applying the above boundary conditions and removing the coupling effect from the tangential components of the electric field, and by defining, $A = \sigma_{xx}k_x + \sigma_{xy}k_y$, $B = \sigma_{yx}k_x + \sigma_{yy}k_y$, $C = \sigma_{xx}k_y - \sigma_{xy}k_x$, and, $D = \sigma_{yy}k_x - \sigma_{yx}k_y$, the unknown coefficients of the TE waves, $a_1 = \frac{\Delta_{TE}^{(1)}}{\Delta_{TE}}$ and $a_1' = \frac{\Delta_{TE}^{(2)}}{\Delta_{TE}}$, can be obtained as [46]:

$$\Delta_{TE} = ik_\rho^2 P^+ Q^+ + P^+ (Ak_x + Bk_y) i\omega \frac{h_1}{k_1} + Q^+ i\omega (Ck_y + Dk_x) + \omega^2 i \frac{h_1}{k_1}(BC + AD) \quad (22)$$

$$\Delta_{TE}^{(1)} = ik_\rho^2 Q^+ P^- + (Ak_x + Bk_y) P^- i\omega \frac{h_1}{k_1} - \omega i (Ck_y + Dk_x) Q^+ - \omega^2 i \frac{h_1}{k_1}(BC + AD) \quad (23)$$

$$\Delta_{TE}^{(2)} = \omega \frac{2h_1}{\mu_1}(Dk_y - Ck_x) \quad (24)$$

where, $P^+ = \frac{h_1}{\mu_1} + \frac{h_2}{\mu_2}$, $Q^+ = \frac{k_1}{\mu_1} + \frac{k_2^2 h_1}{k_1 h_2 \mu_2}$, $P^- = \frac{h_1}{\mu_1} - \frac{h_2}{\mu_2}$, and $Q^- = \frac{k_1}{\mu_1} - \frac{k_2^2 h_1}{k_1 h_2 \mu_2}$. By defining the TM waves expansion coefficients as $b_1 = \frac{\Delta_{TM}^{(1)}}{\Delta_{TM}}$ and $b_1' = \frac{\Delta_{TM}^{(2)}}{\Delta_{TM}}$, it can be shown:

$$\Delta_{TM} = ik_\rho^2 Q^+ P^+ + i\omega Q^+ (Ck_y + Dk_x) + i\omega \frac{h_1}{k_1} P^+ (Ak_x + Bk_y) + i\omega^2 \frac{h_1}{k_1}(AD + BC) \quad (25)$$

$$\Delta_{TM}^{(1)} = -ik_\rho^2 Q^- P^+ + i\omega \frac{h_1}{k_1} P^+ + (Ak_x + Bk_y) - i\omega (Dk_x + Ck_y) Q^- + i\omega^2 \frac{h_1}{k_1}(AD + BC) \quad (26)$$

$$\Delta_{TM}^{(2)} = -2\omega \frac{h_1}{\mu_1}(Ak_y - Bk_x) \quad (27)$$

Other unknown coefficients can be obtained using decoupling equations in [46]. To validate the obtained coefficients, the structure of **Figure 1(a)** consisting of square patches with D_0 = 5 μm, g = 0.5 μm, μ_c = 0.5 eV and τ = 0.5 ps with plane wave illumination is considered under electric and magnetic biases. Since Green's function coefficients are the same as reflection and transmission coefficients of the plane wave, the results of Green's function are compared with the results of the circuit model as [54]:

$$S_{31} = \frac{2(2 + \eta_0 \sigma_d)}{(2 + \eta_0 \sigma_d)^2 + (\eta_0 \sigma_0)^2} \quad (28)$$

$$S_{41} = \frac{2\eta_0 \sigma_0}{(2 + \eta_0 \sigma_d)^2 + (\eta_0 \sigma_0)^2} \quad (29)$$

Figure 3 shows the magnitude of the transmission coefficient by considering the electric bias for the graphene layer. The results of the two methods are identical,

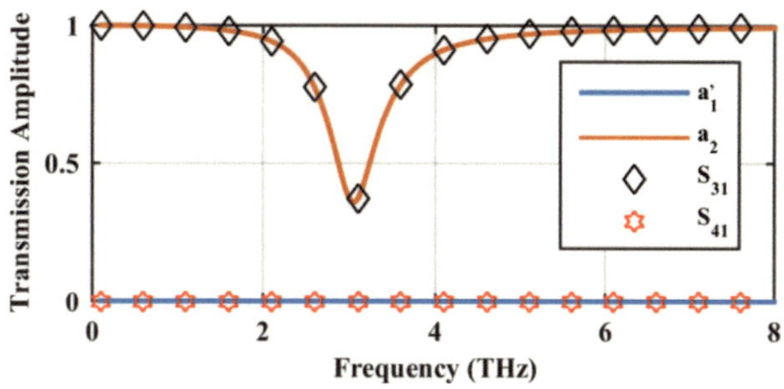

Figure 3.
The magnitude of the transmission coefficient for the graphene nano-patch with the parameters D_o = 5 μm, g = 0.5 μm, μ_c = 0.5 eV, *and* τ = 0.5 ps [46].

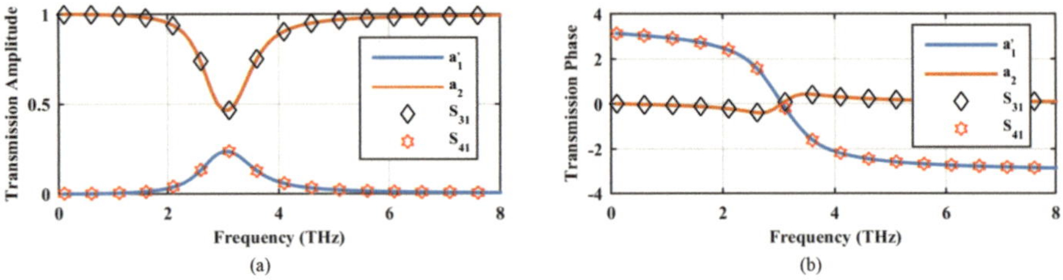

Figure 4.
(a) The magnitude and (b) phase of the transmission coefficient for the graphene nano-patch with $D_o = 5\ \mu m$, $g = 0.5\ \mu m$, $\mu_c = 0.5\ eV$, *ps.* $\tau = 0.5$, *and* $B_o = 0.5\ T$ *[46].*

and because of the absence of electromagnetic coupling under electric bias, the transmission coefficient due to mutual coupling is zero. In **Figure 4** the same results are illustrated for the applied magnetic bias of 0.5 Tesla. There is good agreement in the magnitude and phase of the transmission coefficient in the abovementioned two methods. Also, by finding the poles of the coefficients, the electromagnetic wave propagation constants for the electrically and magnetically biased graphene sheets can be obtained which are in full agreement with [28, 57], respectively. The correctness of the extracted coefficients confirms the validity of the dyadic Green's function formulation.

3.1.2 Graphene-dielectric stack

Dyadic Green's function for an N-layer dielectric environment has been previously formulated using the scattering superposition method [58]. In this section, the above equations are extended to the environment with the electrically biased graphene boundaries, as shown in **Figure 2(b)**. The graphene boundary can be either continuous or periodically patterned as discussed in **section 2**. To start the analysis, the layers are numbered by starting from the top layer, and an arbitrary field point i and source point j are assumed. The problem is solved in the cylindrical coordinates. Since the cylindrical wave functions are discussed in detail in the next section, they are not mentioned here. The dyadic Green's function can be expanded as [58]:

$$\begin{gathered}\overline{\overline{\mathbf{G}}}_{es}^{(ij)}(\boldsymbol{r},\boldsymbol{r}') = \frac{i}{4\pi}\int_{-\infty}^{\infty} dh \sum_{n=0}^{\infty} \frac{(2-\delta_n^0)}{\lambda h_j} \\ \{(1-\delta_i^N)\overline{\boldsymbol{M}}_{n\lambda}(h_i)\left[\left(1-\delta_j^1\right)A_M^{ij}\overline{\boldsymbol{M}}'_{n\lambda}(-h_j)+\left(1-\delta_j^N\right)B_M^{ij}\overline{\boldsymbol{M}}'_{n\lambda}(h_j)\right]+ \\ (1-\delta_i^N)\overline{\boldsymbol{N}}_{n\lambda}(h_i)\left[\left(1-\delta_j^1\right)A_N^{ij}\overline{\boldsymbol{N}}'_{n\lambda}(-h_j)+\left(1-\delta_j^N\right)B_N^{ij}\overline{\boldsymbol{N}}'_{n\lambda}(h_j)\right]+ \\ (1-\delta_i^1)\overline{\boldsymbol{M}}_{n\eta}(-h_i)\left[\left(1-\delta_j^1\right)C_M^{ij}\overline{\boldsymbol{M}}'_{n\lambda}(-h_j)+\left(1-\delta_j^N\right)D_M^{ij}\overline{\boldsymbol{M}}'_{n\lambda}(h_j)\right]+ \\ (1-\delta_i^1)\overline{\boldsymbol{N}}_{n\eta}(-h_i)\times\left[\left(1-\delta_j^1\right)C_N^{ij}\overline{\boldsymbol{N}}'_{n\lambda}(-h_j)+\left(1-\delta_j^N\right)D_N^{ij}\overline{\boldsymbol{N}}'_{n\lambda}(h_j)\right]\}\end{gathered} \tag{30}$$

In the above equations, $\overline{\boldsymbol{M}}$ and $\overline{\boldsymbol{N}}$ are vector wave functions in the cylindrical coordinate system, and $A_{M,N}^{ij}$, $B_{M,N}^{ij}$, $C_{M,N}^{ij}$, $D_{M,N}^{ij}$ are unknown coefficients. The Kranoker delta function is used in the field expansion to generalize the formulation for the arbitrary locations of the field and source points. By applying the boundary conditions on the tangential components of the electric field at the interface of the arbitrary two layers, denoted by i and $i+1$, it can be shown that:

$$\begin{bmatrix} A_M^{ij} \\ B_M^{ij} \\ O_i A_N^{ij} \\ O_i B_N^{ij} \end{bmatrix} e^{ih_i z_i} + \begin{bmatrix} C_M^{ij} \\ D_M^{ij} + \delta_j^i \\ O_i C_N^{ij} \\ O_i\left(D_N^{ij} + \delta_j^i\right) \end{bmatrix} e^{-ih_i z_i} = \begin{bmatrix} \left(A_M^{(i+1)j} + \delta_{i+1}^j\right) \\ B_M^{(i+1)j} \\ O_{i+1}\left(A_N^{(i+1)j} + \delta_{i+1}^j\right) \\ O_{i+1} B_N^{(i+1)j} \end{bmatrix} e^{ih_{i+1} z_i} + \begin{bmatrix} C_M^{(i+1)j} \\ D_M^{(i+1)j} \\ O_{i+1} C_N^{(i+1)j} \\ O_{i+1} D_N^{(i+1)j} \end{bmatrix} e^{-ih_{i+1} z_i} \tag{31}$$

where $O_i = \frac{h_i}{k_i}, P_i = \frac{h_i}{\mu_i}, Q_i = \frac{k_i}{\mu_i}$. The boundary condition on the tangential components of the magnetic field yields:

$$\begin{bmatrix} P_{i+1}\left(A_M^{(i+1)j} + \delta_{i+1}^j\right) \\ P_{i+1} B_M^{(i+1)j} \\ Q_{i+1}\left(A_N^{(i+1)j} + \delta_{i+1}^j\right) \\ Q_{i+1} B_N^{(i+1)j} \end{bmatrix} e^{ih_{i+1} z_i} - \begin{bmatrix} P_{i+1} C_M^{(i+1)j} \\ P_{i+1} D_M^{(i+1)j} \\ Q_{i+1} C_N^{(i+1)j} \\ Q_{i+1} D_N^{(i+1)j} \end{bmatrix} e^{-ih_{i+1} z_i} - \begin{bmatrix} P_i A_M^{ij} \\ P_i B_M^{ij} \\ Q_i A_N^{ij} \\ Q_i B_N^{ij} \end{bmatrix} e^{ih_i z_i}$$
$$+ \begin{bmatrix} P_i C_M^{ij} \\ P_i\left(D_M^{ij} + \delta_j^i\right) \\ Q_i C_N^{ij} \\ Q_i\left(D_N^{ij} + \delta_j^i\right) \end{bmatrix} e^{-ih_i z_i} = i\omega\sigma_{(i+1)i} \begin{bmatrix} A_M^{ij} \\ B_M^{ij} \\ P_i A_N^{ij} \\ P_i B_N^{ij} \end{bmatrix} e^{ih_i z_i} + i\omega\sigma_{(i+1)i} \begin{bmatrix} C_M^{ij} \\ \left(D_M^{ij} + \delta_j^i\right) \\ P_i C_N^{ij} \\ P_i\left(D_N^{ij} + \delta_j^i\right) \end{bmatrix} e^{-ih_i z_i} \tag{32}$$

By re-writing the coefficients as a matrix:

$$\begin{bmatrix} A_{M,N}^{(i+1)j} + \delta_{i+1}^j & B_{M,N}^{(i+1)j} \\ C_{M,N}^{(i+1)j} & D_{M,N}^{(i+1)j} \end{bmatrix} = \begin{bmatrix} \frac{1}{T_{Fi}^{H,V}} & \frac{R_{Fi}^{H,V}}{T_{Fi}^{H,V}} \\ \frac{R_{Pi}^{H,V}}{T_{Pi}^{H,V}} & \frac{1}{T_{Pi}^{H,V}} \end{bmatrix} \begin{bmatrix} A_{M,N}^{ij} & B_{M,N}^{ij} \\ C_{M,N}^{ij} & D_{M,N}^{ij} + \delta_i^j \end{bmatrix} \tag{33}$$

the outgoing and incoming reflection and transmission coefficients can be defined and used to extract the recursive relations. The coefficients for $\overline{\boldsymbol{M}}'$ sources are:

$$R_{Fi}^H = \frac{\mu_i h_{i+1} - \mu_{i+1} h_i + g}{\mu_i h_{i+1} + \mu_{i+1} h_i + g} \tag{34}$$

$$R_{Pi}^H = \frac{\mu_i h_{i+1} - \mu_{i+1} h_i - g}{\mu_i h_{i+1} + \mu_{i+1} h_i - g} \tag{35}$$

$$T_{Fi}^H = \frac{2\mu_i h_{i+1}}{\mu_i h_{i+1} + \mu_{i+1} h_i + g} \tag{36}$$

$$T_{Pi}^{H} = \frac{2\mu_i h_{i+1}}{\mu_i h_{i+1} + \mu_{i+1} h_i - g} \tag{37}$$

The coefficients for $\overline{\mathbf{N}}'$ sources are:

$$R_{Fi}^{V} = \frac{\mu_i h_i k_{i+1}^2 - k_i^2 \mu_{i+1} h_{i+1} + g h_i h_{i+1}}{\mu_i h_i k_{i+1}^2 + \mu_{i+1} h_{i+1} k_i^2 + g h_i h_{i+1}} \tag{38}$$

$$R_{Pi}^{V} = \frac{\mu_i h_i k_{i+1}^2 - \mu_{i+1} k_i^2 h_{i+1} - g h_i h_{i+1}}{\mu_i h_i k_{i+1}^2 + \mu_{i+1} k_i^2 h_{i+1} - g h_i h_{i+1}} \tag{39}$$

$$T_{Pi}^{V} = \frac{2\mu_{i+1} k_i k_{i+1} h_{i+1}}{\mu_i h_i k_{i+1}^2 + \mu_{i+1} k_i^2 h_{i+1} - g h_i h_{i+1}} \tag{40}$$

$$T_{Fi}^{V} = \frac{2\mu_i h_{i+1} k_i k_{i+1}}{\mu_i h_i k_{i+1}^2 + \mu_{i+1} h_{i+1} k_i^2 + g h_i h_{i+1}} \tag{41}$$

The superscripts H and V respectively denote TE and TM sources. Also, subscripts F and P are used to show the incoming and outgoing waves, respectively. The procedure of extracting the unknown coefficients using (34)-(41) is discussed in [58]. To validate the proposed formulas, a parallel plate waveguide with graphene walls is considered. To extract the characteristic equation using the proposed formulations, it is necessary to force the denominator of the coefficients equal to zero. For this three-layer medium:

$$T^{(1)} = T^2.T^1 = .\begin{bmatrix} \frac{1}{T_{F2}^{H}} e^{i(h_2-h_1)d} & \frac{R_{F2}^{H}}{T_{F2}^{H}} e^{-i(h_2-h_1)d} \\ \frac{R_{P2}^{H}}{T_{P2}^{H}} e^{i(h_2+h_1)d} & \frac{1}{T_{P2}^{H}} e^{-i(h_2-h_1)d} \end{bmatrix} \cdot \begin{bmatrix} \frac{1}{T_{F1}^{H}} & \frac{R_{F1}^{H}}{T_{F1}^{H}} \\ \frac{R_{P1}^{H}}{T_{P1}^{H}} & \frac{1}{T_{P1}^{H}} \end{bmatrix} \tag{42}$$

From which it can be deduced that $B_{M,N}^{11} = -\frac{T_{12}^{(1)}}{T_{11}^{(1)}}$. By setting $T_{11}^{(1)}$ equal to zero and assuming that the medium (1) and (3) are the same, and also defining $h = i\omega\mu_1\sigma$, for the H coefficients it can be concluded that:

$$\frac{e^{ih_2 d} - 1}{e^{ih_2 d} + 1} h_1 - (h_2 + h) = 0 \ \Rightarrow h_2 + h + j\tan\left(\frac{h_2 d}{2}\right) h_1 = 0 \tag{43}$$

This procedure is repeatable for V sources. Also, to calculate the reflection coefficient from a multilayer structure, it is necessary to consider the field and source points in region 1. In this case, the only non-zero coefficient in Green's function expansion is $B_{M,N}^{11}$ coefficient representing the plane wave reflection coefficient from the multilayer structure.

3.2 Graphene-based cylindrical structures

In this section, the dyadic Green's function of a cylindrical structure with the tensor surface conductivity boundary condition will be extracted. Later, different examples of guiding and scattering problems are provided to investigate the validity of the formulation. In general, in cylindrical structures, TE and TM modes are coupled, which leads to the complexity of mathematical relations. Therefore, the generalization of the formulation to the multilayered cylinders is not considered here. Note that graphene sheets can be wrapped around cylindrical particles due to the presence of van der Waals force [59]. For this purpose, tape-assist transfer under micromanipulation and spin-coating methods are proposed [60].

3.2.1 Dyadic Green's function for a cylinder with tensor surface conductivity boundary condition

The dyadic Green's function of the single-layer cylinder with the tensor surface conductivity boundary condition, as considered in **Figure 5.**, will be extracted in the following. The interior region of the cylinder is made of dielectric material and its cover is considered as a full tensor surface conductivity. To solve the problem, the vector wave functions are defined as [23]:

$$\overline{\boldsymbol{M}}_\mu(h) = e^{im\phi}e^{ihz}\left[im\frac{Z_m(\mu r)}{r}\hat{\boldsymbol{r}} - \frac{\partial Z_m(\mu r)}{\partial r}\hat{\boldsymbol{\phi}}\right] \tag{44}$$

$$\overline{\boldsymbol{N}}_\mu(h) = \frac{1}{k_j}e^{im\phi}e^{ihz}\left[ih\frac{\partial Z_m(\mu r)}{\partial r}\hat{\boldsymbol{r}} - \frac{hm}{r}Z_m(\mu r)\hat{\boldsymbol{\phi}} - \mu^2 Z_m(\mu r)\hat{\boldsymbol{z}}\right] \tag{45}$$

In the above equations, Z_m is the cylindrical Bessel function in the inner layer and the cylindrical Hankel function in the outer layer, both with the orders of m. The wavenumber in the radial direction is μ and the wavenumber along the length is h. The free-space Green's function in the cylindrical coordinates is [23]:

$$\begin{aligned}\overline{\overline{\mathbf{G}}}_{e0}\left(\overline{\boldsymbol{R}},\overline{\boldsymbol{R}}'\right) &= -\frac{1}{k^2}\hat{\boldsymbol{r}}\hat{\boldsymbol{r}}\delta\left(\overline{\boldsymbol{R}}-\overline{\boldsymbol{R}}'\right)\\ &+\frac{i}{8\pi}\int_{-\infty}^{\infty}dh\sum_{m=-\infty}^{\infty}\frac{1}{\eta^2}\times\left[\overline{\boldsymbol{M}}_\eta(h)\overline{\boldsymbol{M}}_\eta'^{(1)}(-h)+\overline{\boldsymbol{N}}_\eta(h)\overline{\boldsymbol{N}}_\eta'^{(1)}(-h)\right]\quad R<R'\end{aligned} \tag{46}$$

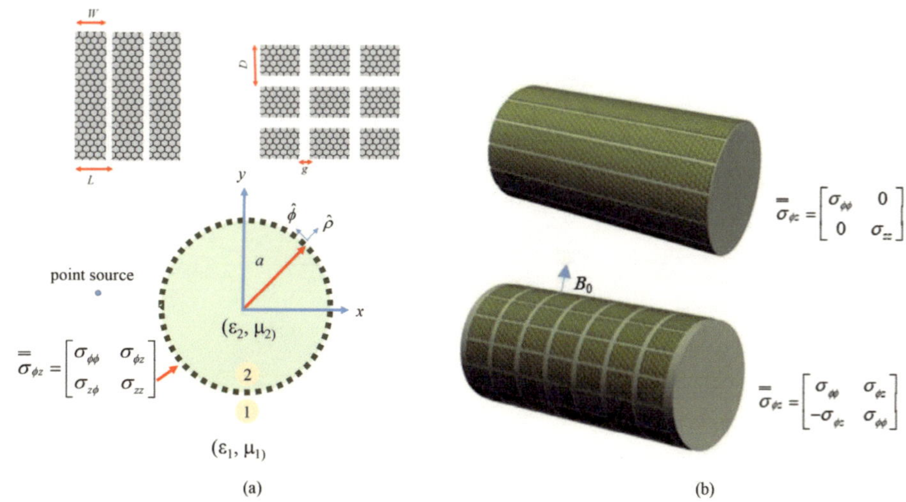

Figure 5.
(a) A monolayer cylinder with a 2D cover with the tensor surface conductivity boundary condition and (b) its special cases constructed by densely packed strips and square nano-patches [61].

In the cylindrical coordinate system, TE and TM modes are coupled and the expansion of the fields are [23].

$$\overline{\overline{G}}_{es}^{(11)}\left(\overline{R},\overline{R}'\right)=\frac{i}{8\pi}\int_{-\infty}^{\infty}dh\sum_{m=-\infty}^{\infty}\frac{1}{\eta^2}\times$$

$$\left\{\left[A_\eta\overline{M}_\eta^{(1)}(h)+B_\eta\overline{N}_\eta^{(1)}(h)\right]\overline{M}_\eta'^{(1)}(-h)+\left[C_\eta\overline{N}_\eta^{(1)}(h)+D_\eta\overline{M}_\eta^{(1)}(h)\right]\overline{N}_\eta'^{(1)}(-h)\right\} \tag{47}$$

$$\overline{\overline{G}}_{es}^{(21)}\left(\overline{R},\overline{R}'\right)=\frac{i}{8\pi}\int_{-\infty}^{\infty}dh\sum_{m=-\infty}^{\infty}\frac{1}{\eta^2}\times \tag{48}$$

$$\left\{\left[a_\xi\overline{M}_\xi(h)+b_\xi\overline{N}_\xi(h)\right]\overline{M}_\eta'^{(1)}(-h)+\left[c_\xi\overline{N}_\xi(h)+d_\xi\overline{M}_\xi(h)\right]\overline{N}_\eta'^{(1)}(-h)\right\}$$

where A_η, B_η, C_η, and D_η are the unknown coefficients of the DGF expansion in region 1. Also, a_ξ, b_ξ, c_ξ, and d_ξ are DGF expansion coefficients in the region 2. The boundary condition on the tangential components of the magnetic Green's function is given by [61]:

$$\hat{r}\times\left(\frac{\nabla\times\overline{\overline{G}}_e^{(11)}}{\mu_1}-\frac{\nabla\times\overline{\overline{G}}_e^{(21)}}{\mu_2}\right)=i\omega\overline{\overline{\sigma}}_{\phi z}.\overline{\overline{G}}_e^{(21)} \tag{49}$$

By applying the above-mentioned boundary condition along with the boundary condition regarding the continuity of the electric Green's function to **(47)–(48)**, the system of equations to determine the unknown coefficients can be obtained as:

$$\begin{bmatrix} -\dfrac{\partial H_m^{(1)}(\eta a)}{\partial a} & -\dfrac{1}{k_1}\dfrac{hm}{a}H_m^{(1)}(\eta a) & \dfrac{\partial J_m(\xi a)}{\partial a} & \dfrac{1}{k_2}\dfrac{hm}{a}J_m(\xi a) \\ 0 & -\dfrac{1}{k_1}\eta^2H_m^{(1)}(\eta a) & 0 & \dfrac{1}{k_2}\xi^2J_m(\xi a) \\ -\dfrac{1}{\mu_1}\dfrac{hm}{a}H_m^{(1)}(\eta a) & -\dfrac{k_1}{\mu_1}\dfrac{\partial H_m^{(1)}(\eta a)}{\partial a} & \dfrac{1}{\mu_2}\dfrac{hm}{a}J_m(\xi a)+\sigma_1 & \dfrac{k_2}{\mu_2}\dfrac{\partial J_m(\xi a)}{\partial a}+\sigma_2 \\ \dfrac{1}{\mu_1}\eta^2H_m^{(1)}(\eta a) & 0 & -\dfrac{1}{\mu_2}\xi^2J_m(\xi a)+\sigma_3 & \sigma_4 \end{bmatrix}\times\begin{bmatrix}A_\eta\\B_\eta\\a_\xi\\b_\xi\end{bmatrix}$$

$$=\begin{bmatrix}\dfrac{\partial J_m(\eta a)}{\partial a}\\0\\\dfrac{1}{\mu_1}\dfrac{hm}{a}J_m(\eta a)\\-\dfrac{1}{\mu_1}\eta^2J_m(\eta a)\end{bmatrix} \tag{50}$$

$$\begin{bmatrix} -\frac{1}{k_1}\frac{hm}{a}H_m^{(1)}(\eta a) & -\frac{\partial H_m^{(1)}(\eta a)}{\partial a} & \frac{1}{k_2}\frac{hm}{a}J_n(\xi a) & \frac{\partial J_m(\xi a)}{\partial a} \\ -\frac{1}{k_1}\eta^2 H_m^{(1)}(\eta a) & 0 & \frac{1}{k_2}\xi^2 J_m(\xi a) & 0 \\ -\frac{k_1}{\mu_1}\frac{\partial H_m^{(1)}(\eta a)}{\partial a} & -\frac{1}{\mu_1}\frac{hm}{a}H_m^{(1)}(\eta a) & \frac{k_2}{\mu_2}\frac{\partial J_m(\xi a)}{\partial a}+\sigma_1' & \frac{1}{\mu_2}\frac{hm}{a}J_m(\xi a)+\sigma_2' \\ 0 & \frac{1}{\mu_1}\eta^2 H_m^{(1)}(\eta a) & \sigma_3' & -\frac{1}{\mu_2}\xi^2 J_m(\xi a)+\sigma_4' \end{bmatrix} \times \begin{bmatrix} C_\eta \\ D_\eta \\ c_\xi \\ d_\xi \end{bmatrix} = \begin{bmatrix} \frac{1}{k_1}\frac{hm}{a}J_m(\eta a) \\ \frac{1}{k_1}\eta^2 J_m(\eta a) \\ \frac{k_1}{\mu_1}\frac{\partial J_m(\eta a)}{\partial a} \\ 0 \end{bmatrix}. \tag{51}$$

where:

$$\sigma_1 = i\omega\sigma_{z\phi}\frac{\partial J_m(\xi a)}{\partial a}, \sigma_2 = i\omega\frac{\sigma_{zz}}{k_2}\xi^2 J_m(\xi a) + i\omega\sigma_{z\phi}\frac{hm}{k_2 a}J_m(\xi a)$$

$$\sigma_3 = i\omega\sigma_{\phi\phi}\frac{\partial J_m(\xi a)}{\partial a}, \sigma_4 = i\omega\sigma_{\phi\phi}\frac{1}{k_2}\frac{hm}{a}J_m(\xi a) + i\omega\sigma_{\phi z}\frac{1}{k_2}\xi^2 J_m(\xi a)$$

$$\sigma_1' = i\omega\sigma_{zz}\frac{1}{k_2}\xi^2 J_m(\xi a) + i\omega\sigma_{z\phi}\frac{hm}{k_2 a}J_m(\xi a), \sigma_2' = i\omega\sigma_{z\phi}\frac{\partial J_m(\xi a)}{\partial a}$$

$$\sigma_3' = i\omega\sigma_{\phi\phi}\frac{1}{k_2}\frac{hm}{a}J_m(\xi a) + i\omega\sigma_{\phi z}\frac{1}{k_2}\xi^2 J_m(\xi a), \sigma_4' = i\omega\sigma_{\phi\phi}\frac{\partial J_m(\xi a)}{a}$$

Nullifying the determinant of the matrices **(50)–(51)** for $m = 0$, each of the above matrices can be separated as the multiplication of:

$$\frac{\omega\varepsilon_2}{\xi}\frac{J_1(\xi a)}{J_0(\xi a)} - \frac{\omega\varepsilon_1}{\eta}\frac{H_1^{(1)}(\eta a)}{H_0^{(1)}(\eta a)} + i\sigma_d = 0 \tag{52}$$

$$\frac{\xi}{\omega\mu_2}\frac{J_0(\xi a)}{J_1(\xi a)} - \frac{\eta}{\omega\mu_1}\frac{H_0^{(1)}(\eta a)}{H_1^{(1)}(\eta a)} + i\sigma_d = 0 \tag{53}$$

For the graphene shell under magnetic bias, both matrices in **(50)–(51)** result in the following equation for the propagation constant of hybrid TE and TM waves:

$$\left[\frac{\omega\varepsilon_2}{\xi}\frac{H_1^{(1)}(-\xi a)}{H_0^{(1)}(-\xi a)} - \frac{\omega\varepsilon_1}{\eta}\frac{J_1(-\eta a)}{J_0(-\eta a)} + i\sigma_d\right] \times \left[\frac{\eta}{\omega\mu_1}\frac{J_0(-\eta a)}{J_1(-\eta a)} - \frac{\xi}{\omega\mu_2}\frac{H_0^{(1)}(-\xi a)}{H_1^{(1)}(-\xi a)} + i\sigma_d\right] = -\sigma_0^2 \tag{54}$$

Which is in agreement with [57]. Also, the total scattering cross-section (TSCS) of a graphene-coated cylinder with the parameters a = 50 μm, ε_2 = 2.4, τ = 1 ps, μ_c = 0.25 eV under electric bias is calculated for both TE and TM polarizations in

Figure 6, and compared with the results of the CST2017 software package. As can be seen, both methods have resulted in the same results.

As another example, the densely packed graphene strips with the parameters L = 420 nm, W = 400 nm, μ_c = 0.5 eV, and τ = 1 ps are considered around the dielectric cylinder as in **Figure 5.** It is assumed that the strips are wrapped around a hollow cylinder with the radius of a = 50 μm. **Figure** 7 shows the TSCS of this structure for the magnetic biases with the strength in the range of 20-40 T. As observed, by increasing the magnetic bias, the resonant frequency of the surface plasmons blue shifts. The associated planar structure behaves as a hyperbolic meta-surface [62]. Under locally flat consideration of the curvature, this structure can also be considered as a hyperbolic medium. In the cylindrical geometries, hyperbolic meta-surfaces can be obtained using graphene-dielectric stacks [63]. The advantage of this hyperbolic structure is its two-dimensional nature and reconfigurability. It is also demonstrated that covering the surface of nanotubes with the hyperbolic meta-surface increases the interaction of the light with dipole emitters [64].

Finally, as **Figure 5(b)** illustrates, graphene-based square patches around the cylinder are considered under magnetic bias. Geometrical and optical parameters are as follows: τ = 1 ps, g = 0.5 μm, and D = 0.5 μm and the TSCS is illustrated in **Figure 8.** for B_0 = 0 T and B_0 = 10 T. As can be seen, by changing the magnetic bias, the optical state changes from the maximum scattering to the minimum scattering. Such capability has recently been proposed by using a phase change material for switching between these two situations [65]. In this structure, the operating frequency can also be adjusted by changing the electric bias of graphene.

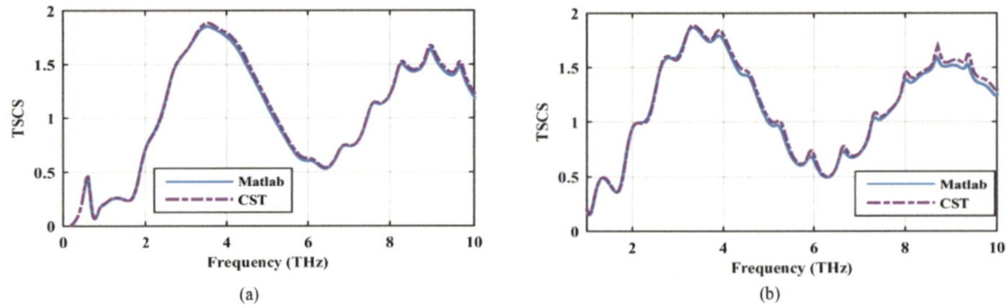

Figure 6.
TSCS for an infinite cylinder covered with electrically biased graphene shell with a = 50 *μm,* ε_2 = 2.4, τ = 1 *ps, and* μ_c = 0.25 *eV considering (a) TE and (b) TM polarization for the incident wave [61].*

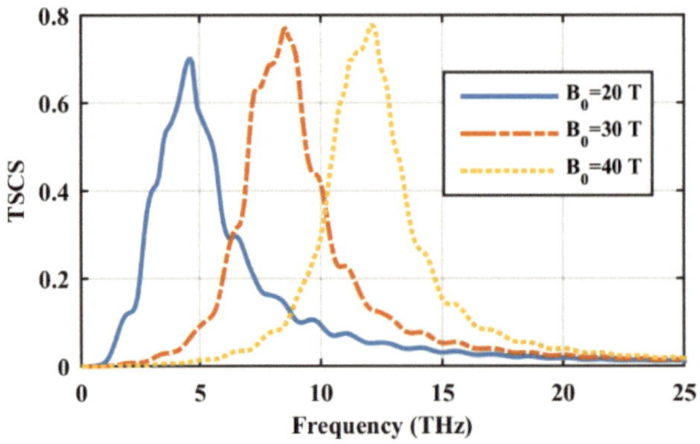

Figure 7.
TSCS for an infinite length hollow cylinder (a = 50 *μm) coated with densely packed graphene strips with* L = 420 *nm,* W = 400 *nm,* μ_c = 0.5 *eV, and* τ = 1 *ps considering different magnetic biases [61].*

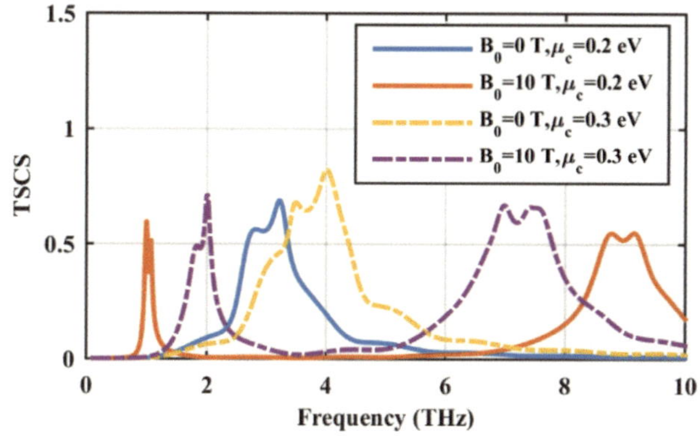

Figure 8.
TSCS for an infinite cylinder coated with graphene patches with D = 0.5 *μm,* g = 0.5 *μm, and* τ = 0.5 *ps under different electric and magnetic biases [61].*

It is essential to note that in the cylindrical structures, the spatial domain Green's function consists of an integral on the real axis. Due to the existence of poles in the integration path, the integration path is usually deformed into a triangular shape. Also, a common method for calculating the spatial domain Green's function is the generalized pencil of function (GPOF) method in which the Green's function is expanded in terms of the complex exponential functions [66]. The unknowns can be found via the algorithm provided in [67]. Also, as mentioned earlier, the dyadic Green's function of the graphene-based multilayered cylindrical structures will not be considered here. For calculating the scattering cross-section of graphene-based multilayer cylindrical structures, a simple approach based on the transfer matrix method (TMM) is proposed that will be suitable for establishing novel optical devices [68, 69].

3.3 Graphene-based spherically layered medium

In this section, a multi-layered spherical structure with the graphene boundaries is considered and its Green's function is extracted by assuming different locations for the source and observation points. The relationship between the Green's function expansion coefficients and the modified Mie-Lorentz coefficients is exhibited to discuss how to solve the scattering problems using the Green's function. Scattering analysis of graphene-based layered structures is of great importance in the design of novel optical devices [70]. Finally, the procedure for calculating the Purcell factor is considered as an important application. Instances of experimentally realized graphene-coated spherical particles can be found in [71, 72], where improved template method and hydrothermal method are proposed for the synthesis. Also, transmission electron microscopy (TEM) and field emission scanning electron microscopy (FE-SEM) are used for characterization.

3.3.1 Dyadic Green's function of a graphene-based spherically layered structure

Let us consider an *N*-layer spherical medium with the graphene boundaries as shown in **Figure 9.** The purpose of this section is to compute the dyadic Green's function of this structure with the assumption of arbitrary locations for the field and source points. For this purpose, the Green's function in each layer is expanded in terms of vector wave functions with unknown coefficients. These functions are calculated using the scalar wave function of [74]:

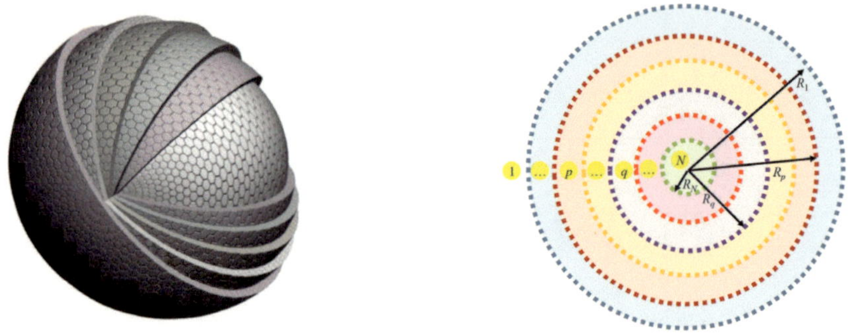

(a) (b)

Figure 9.
Spherically layered medium with graphene boundaries (a) 2D and (b) 3D views [73].

$$\varphi_{mn}(r,\theta,\phi) = z_n(k_p r) P_n^m(\cos\theta) e^{im\phi} \tag{55}$$

where $z_n(.)$ represents the spherical Bessel or Hankel functions of order n and $P_n^m(.)$ is the associated Legendre function with degree n and order m. It can be readily shown that:

$$\overline{\boldsymbol{M}}_{mn}(k_p) = z_n(k_p r)\, e^{im\phi} \left[\frac{im}{\sin\theta} P_n^m(\cos\theta)\hat{\boldsymbol{\theta}} - \frac{dP_n^m(\cos\theta)}{d\theta}\hat{\boldsymbol{\phi}} \right] \tag{56}$$

$$\begin{aligned} \overline{\boldsymbol{N}}_{mn}(k_p) &= \frac{n(n+1)}{k_p r} z_n(k_p r) P_n^m(\cos\theta)\, e^{im\phi}\hat{\boldsymbol{r}} + \\ \frac{1}{k_p r}\frac{d\left[r z_n(k_p r)\right]}{dr} &\left[\frac{dP_n^m(\cos\theta)}{d\theta}\hat{\boldsymbol{\theta}} + \frac{im}{\sin\theta} P_n^m(\cos\theta)\hat{\boldsymbol{\phi}} \right] e^{im\phi} \end{aligned} \tag{57}$$

The above vector functions are self and mutually orthogonal. This feature will be used to decouple the equation when computing the unknown coefficients.

As mentioned earlier, in the scattering superposition method, the dyadic Green's function is written as the sum of the free-space and scattering Green's functions. The free-space Green's function is related to the source in an infinite homogeneous medium while the scattering Green's function is due to the source in the presence of the layered medium. The expansion of the Green's function for the spherical structure with N concentric layers, assuming that the source and field points are respectively located in the desired layers with the labels p and q, can be written as [23]:

$$\overline{\overline{\hat{\boldsymbol{G}}}}_e^{(pq)}(\boldsymbol{r},\boldsymbol{r}') = \overline{\overline{\hat{\boldsymbol{G}}}}_{0e}^{(pq)}(\boldsymbol{r},\boldsymbol{r}')\delta_q^p + \overline{\overline{\hat{\boldsymbol{G}}}}_{es}^{(pq)}(\boldsymbol{r},\boldsymbol{r}') \tag{58}$$

The free-space Green's function can be obtained using the residue theorem as [23]:

$$\begin{aligned} \overline{\overline{\hat{\boldsymbol{G}}}}_{0e}(\boldsymbol{r},\boldsymbol{r}') &= -\frac{\hat{\boldsymbol{r}}\hat{\boldsymbol{r}}}{k_q^2}\delta(\boldsymbol{r}-\boldsymbol{r}') + \frac{ik_q}{4\pi}\sum_{n=1}^{\infty}\sum_{m=-n}^{n}\left(2-\delta_m^0\right)\frac{2n+1}{n(n+1)} \times \\ \frac{(n-m)!}{(n+m)!} &\times \begin{cases} \overline{\boldsymbol{M}}_{mn}^{(1)}(k_q)\overline{\boldsymbol{M}}'_{mn}(k_q) + \overline{\boldsymbol{N}}_{mn}^{(1)}(k_q)\overline{\boldsymbol{N}}'_{mn}(k_q) & r > r' \\ \overline{\boldsymbol{M}}_{mn}(k_q)\overline{\boldsymbol{M}}'^{(1)}_{mn}(k_q) + \overline{\boldsymbol{N}}_{mn}(k_q)\overline{\boldsymbol{N}}'^{(1)}_{mn}(k_q) & r < r' \end{cases} \end{aligned} \tag{59}$$

Moreover, scattering Green's function in each layer can be expanded as [75]:

$$\hat{\overline{\overline{G}}}_{es}^{(pq)}(\boldsymbol{r},\boldsymbol{r}') = \frac{ik_q}{4\pi}\sum_{n=1}^{\infty}\sum_{m=-n}^{n}\left(2-\delta_m^0\right)\frac{2n+1}{n(n+1)}\frac{(n-m)!}{(n+m)!}\times$$
$$\left\{\left(1-\delta_p^N\right)\overline{\boldsymbol{M}}_{mn}^{(1)}\left[\left(1-\delta_q^1\right)A_H^{pq}\overline{\boldsymbol{M}}'_{mn}+\left(1-\delta_q^N\right)B_H^{pq}\overline{\boldsymbol{M}}'^{(1)}_{mn}\right]\right.$$
$$+\left(1-\delta_p^N\right)\overline{\boldsymbol{N}}_{mn}^{(1)}\left[\left(1-\delta_q^1\right)A_V^{pq}\overline{\boldsymbol{N}}'_{mn}+\left(1-\delta_q^N\right)B_V^{pq}\overline{\boldsymbol{N}}'^{(1)}_{mn}\right] \tag{60}$$
$$+\left(1-\delta_p^1\right)\overline{\boldsymbol{M}}_{mn}\left[\left(1-\delta_q^1\right)C_H^{pq}\overline{\boldsymbol{M}}'_{mn}+\left(1-\delta_q^N\right)D_H^{pq}\overline{\boldsymbol{M}}'^{(1)}_{mn}\right]$$
$$\left.+\left(1-\delta_p^1\right)\overline{\boldsymbol{N}}_{mn}\left[\left(1-\delta_q^1\right)C_V^{pq}\overline{\boldsymbol{N}}'_{mn}+\left(1-\delta_q^N\right)D_V^{pq}\overline{\boldsymbol{N}}'^{(1)}_{mn}\right]\right\}$$

where $A_{H,V}^{pq}$, $B_{H,V}^{pq}$, $C_{H,V}^{pq}$, and $D_{H,V}^{pq}$ are the unknown coefficients of the Green's function to be obtained. In the above formulation, the superscript (1) represents that the spherical Hankel functions of the first type are chosen to represent the spherical vector functions. For the other vector wave functions, the first-kind Bessel function should be selected. As observed, in the spherical structures the TE and TM waves are not coupled. Thus, the expansion of the fields only includes the interaction of the functions $\overline{\boldsymbol{M}}$ and $\overline{\boldsymbol{M}}'$ as well as the functions $\overline{\boldsymbol{N}}$ and $\overline{\boldsymbol{N}}'$. Moreover, to solve the problem for the structures with the arbitrary number of layers, the Kronecker delta function is used in the expansions of the fields. In the middle layers of the spherical structure, the electric field is a linear combination of the Bessel and Hankel functions, whereas in the outermost layer, only the Hankel functions, and in the innermost layer, only the Bessel functions will exist. So we have used $\left(1-\delta_p^1\right)$ and $\left(1-\delta_p^N\right)$ functions. To determine the delta functions related to the source terms, free-space Green's function can be used. For $q = 1$ (external layer), the source functions related to the second criterion of the free-space Green's function is used. Also, for $q = N$ (internal layer) the source functions related to the first criterion should be used. In the middle layers, a linear combination of different source functions must be used.

It should be noted that by using the addition theorem in Legendre functions, the internal series in the Green's function expansion can be eliminated [11]. To select the number of terms required for the convergence of the external series, the conditions of the problem must be considered. In the other words, in the structures whose electrical size is very much smaller than that of the wavelength, the term $n = 1$ is sufficient for the convergence [76]. Also, in the case where the distance between the source and observation points is large, the series can be truncated in the number $N_t = x + 3\sqrt{x} + 2$, where $x = k_0R_1$. Otherwise, the convergence of the series is weak, and a large number of terms in the range of 20 k_0R_1 should be considered. In this case, the series acceleration techniques will be highly efficient in terms of computational efficiency [77, 78].

Boundary conditions on tangential components of electric and magnetic Green's functions in the interface of two adjacent layers are [73]:

$$\hat{\boldsymbol{r}}\times\hat{\overline{\overline{G}}}_{e}^{(pq)} = \hat{\boldsymbol{r}}\times\hat{\overline{\overline{G}}}_{e}^{[(p+1)q]} \tag{61}$$

$$\frac{1}{i\omega\mu_{p+1}}\hat{\boldsymbol{r}}\times\nabla\times\hat{\overline{\overline{G}}}_{e}^{[(p+1)q]} - \frac{1}{i\omega\mu_{p}}\hat{\boldsymbol{r}}\times\nabla\times\hat{\overline{\overline{G}}}_{e}^{(pq)} = -\sigma_{(p+1)p}\hat{\boldsymbol{r}}\times\left(\hat{\boldsymbol{r}}\times\hat{\overline{\overline{G}}}_{e}^{(pq)}\right) \tag{62}$$

which respectively lead to the following equations:

$$\begin{bmatrix} \xi_n^{pp} A_H^{pq} \\ \xi_n^{pp} B_H^{pq} \\ \partial\xi_n^{pp} A_V^{pq} \\ \partial\xi_n^{pp} B_V^{pq} \end{bmatrix} + \begin{bmatrix} \psi_n^{pp} C_H^{pq} \\ \psi_n^{pp} \left(D_H^{pq} + \delta_p^q\right) \\ \partial\psi_n^{pp} C_V^{pq} \\ \partial\psi_n^{pp} \left(D_V^{pq} + \delta_p^q\right) \end{bmatrix} = \begin{bmatrix} \xi_n^{(p+1)p} \left(A_H^{(p+1)q} + \delta_{p+1}^q\right) \\ \xi_n^{(p+1)p} B_H^{(p+1)q} \\ \partial\xi_n^{(p+1)p} \left(A_V^{(p+1)q} + \delta_{p+1}^q\right) \\ \partial\xi_n^{(p+1)p} B_V^{(p+1)q} \end{bmatrix} + \begin{bmatrix} \psi_n^{(p+1)p} C_H^{(p+1)q} \\ \psi_n^{(p+1)p} D_H^{(p+1)q} \\ \partial\psi_n^{(p+1)p} C_V^{(p+1)q} \\ \partial\psi_n^{(p+1)q} D_V^{(p+1)q} \end{bmatrix} \tag{63}$$

$$\frac{k_{p+1}}{\mu_{p+1}} \begin{bmatrix} \partial\xi_n^{(p+1)p} \left(A_H^{(p+1)p} + \delta_{p+1}^q\right) \\ \partial\xi_n^{(p+1)p} B_H^{(p+1)p} \\ \xi_n^{(p+1)p} \left(A_V^{(p+1)q} + \delta_{p+1}^q\right) \\ \xi_n^{(p+1)p} B_V^{(p+1)q} \end{bmatrix} + \frac{k_{p+1}}{\mu_{p+1}} \begin{bmatrix} \partial\psi_n^{(p+1)p} C_H^{(p+1)q} \\ \partial\psi_n^{(p+1)p} D_H^{(p+1)q} \\ \psi_n^{(p+1)p} C_V^{(p+1)q} \\ \psi_n^{(p+1)p} D_V^{(p+1)q} \end{bmatrix} - \frac{k_p}{\mu_p} \begin{bmatrix} \partial\xi_n^{pp} A_H^{pq} \\ \partial\xi_n^{pp} B_H^{pq} \\ \xi_n^{pp} A_V^{pq} \\ \xi_n^{pp} B_V^{pq} \end{bmatrix} - \frac{k_p}{\mu_p} \begin{bmatrix} \partial\psi_n^{pp} C_H^{pq} \\ \partial\psi_n^{pp} \left(D_H^{pq} + \delta_p^q\right) \\ \psi_n^{pp} C_V^{pq} \\ \psi_n^{pp} \left(D_V^{pq} + \delta_p^q\right) \end{bmatrix} = \sigma_{(p+1)p} \begin{bmatrix} \xi_n^{pp} A_H^{pq} \\ \xi_n^{pp} B_H^{pq} \\ \partial\xi_n^{pp} A_V^{pq} \\ \partial\xi_n^{pp} B_V^{pq} \end{bmatrix} + \sigma_{(p+1)p} \begin{bmatrix} \psi_n^{pp} C_H^{pq} \\ \psi_n^{pp} \left(D_H^{pq} + \delta_p^q\right) \\ \partial\psi_n^{pp} C_V^{pq} \\ \partial\psi_n^{pp} \left(D_V^{pq} + \delta_p^q\right) \end{bmatrix} \tag{64}$$

In the above equations, $\psi_n^{pq} = j_n\left(k_p R_q\right)$, $\xi_n^{pq} = h_n^{(1)}\left(k_p R_q\right)$, $\partial\psi_n^{pq} = 1/\rho\, \mathrm{d}\left[\rho j_n(\rho)\right]\big|_{\rho=k_p R_q}$ and $\partial\xi_n^{pq} = 1/\rho\, \mathrm{d}\left[\rho h_n^{(1)}(\rho)\right]\Big|_{\rho=k_p R_q}$. To obtain recursive relations for unknown coefficients of Green's function, separating the above equations is necessary. Therefore [75]:

$$\begin{bmatrix} A_{H,V}^{(p+1)q} + \delta_{p+1}^q \\ B_{H,V}^{(p+1)q} \end{bmatrix} = \frac{1}{T_{Fp}^{H,V}} \begin{bmatrix} A_{H,V}^{pq} \\ B_{H,V}^{pq} \end{bmatrix} + \frac{R_{Fp}^{H,V}}{T_{Fp}^{H,V}} \begin{bmatrix} C_{H,V}^{pq} \\ D_{H,V}^{pq} + \delta_p^q \end{bmatrix} \tag{65}$$

$$\begin{bmatrix} C_{H,V}^{(p+1)q} \\ D_{H,V}^{(p+1)q} \end{bmatrix} = \frac{R_{Pp}^{H,V}}{T_{Pp}^{H,V}} \begin{bmatrix} A_{H,V}^{pq} \\ B_{H,V}^{pq} \end{bmatrix} + \frac{1}{T_{Pp}^{H,V}} \begin{bmatrix} C_{H,V}^{pq} \\ D_{H,V}^{pq} + \delta_p^q \end{bmatrix} \tag{66}$$

where the reflection and transmission coefficients for TE waves are computed as:

$$R_{Fp}^H = \frac{k_{p+1}\mu_p \partial\psi_n^{(p+1)p} \psi_n^{pp} - k_p \mu_{p+1} \partial\psi_n^{pp} \psi_n^{(p+1)p} + g\, \psi_n^{pp} \psi_n^{(p+1)p}}{k_{p+1}\mu_p \partial\psi_n^{(p+1)p} \xi_n^{pp} - k_p \mu_{p+1} \partial\xi_n^{pp} \psi_n^{(p+1)p} + g\, \xi_n^{pp} \psi_n^{(p+1)p}} \tag{67}$$

$$R_{Pp}^H = \frac{k_{p+1}\mu_p \partial\xi_n^{(p+1)p} \xi_n^{pp} - k_p \mu_{p+1} \partial\xi_n^{pp} \xi_n^{(p+1)p} + g\, \xi_n^{pp} \xi_n^{(p+1)p}}{k_{p+1}\mu_p \partial\xi_n^{(p+1)p} \psi_n^{pp} - k_p \mu_{p+1} \partial\psi_n^{pp} \xi_n^{(p+1)p} + g\, \psi_n^{pp} \xi_n^{(p+1)p}} \tag{68}$$

$$T_{Fp}^{H}=\frac{k_{p+1}\mu_{p}\left(\partial\psi_{n}^{(p+1)p}\xi_{n}^{(p+1)p}-\psi_{n}^{(p+1)p}\partial\xi_{n}^{(p+1)p}\right)}{k_{p+1}\mu_{q}\partial\psi_{n}^{(p+1)p}\xi_{n}^{pp}-k_{p}\mu_{p+1}\partial\xi_{n}^{pp}\psi_{n}^{(p+1)p}+g\,\xi_{n}^{pp}\psi_{n}^{(p+1)p}}\tag{69}$$

$$T_{Pp}^{H}=\frac{k_{p+1}\mu_{p}\left(\partial\xi_{n}^{(p+1)p}\psi_{n}^{(p+1)p}-\partial\psi_{n}^{(p+1)p}\xi_{n}^{(p+1)p}\right)}{k_{p+1}\mu_{p}\partial\xi_{n}^{(p+1)p}\psi_{n}^{pp}-k_{p}\mu_{p+1}\partial\psi_{n}^{pp}\xi_{n}^{(p+1)p}+g\,\psi_{n}^{pp}\xi_{n}^{(p+1)p}}\tag{70}$$

Also, the reflection and transmission coefficients for the TM waves are:

$$R_{Fp}^{V}=\frac{k_{p+1}\mu_{p}\psi_{n}^{(p+1)p}\partial\psi_{n}^{pp}-k_{p}\mu_{p+1}\psi_{n}^{pp}\partial\psi_{n}^{(p+1)p}+g\,\partial\psi_{n}^{pp}\partial\psi_{n}^{(p+1)p}}{k_{p+1}\mu_{p}\psi_{n}^{(p+1)p}\partial\xi_{n}^{pp}-k_{p}\mu_{p+1}\xi_{n}^{pp}\partial\psi_{n}^{(p+1)p}+g\,\partial\xi_{n}^{pp}\partial\psi_{n}^{(p+1)p}}\tag{71}$$

$$R_{Fp}^{V}=\frac{k_{p+1}\mu_{p}\psi_{n}^{(p+1)p}\partial\psi_{n}^{pp}-k_{p}\mu_{p+1}\psi_{n}^{pp}\partial\psi_{n}^{(p+1)p}+g\,\partial\psi_{n}^{pp}\partial\psi_{n}^{(p+1)p}}{k_{p+1}\mu_{p}\psi_{n}^{(p+1)p}\partial\xi_{n}^{pp}-k_{p}\mu_{p+1}\xi_{n}^{pp}\partial\psi_{n}^{(p+1)p}+g\,\partial\xi_{n}^{pp}\partial\psi_{n}^{(p+1)p}}\tag{72}$$

$$T_{Fp}^{V}=\frac{k_{p+1}\mu_{p}\left(\psi_{n}^{(p+1)p}\partial\xi_{n}^{(p+1)p}-\xi_{n}^{(p+1)p}\partial\psi_{n}^{(p+1)p}\right)}{k_{p+1}\mu_{p}\psi_{n}^{(p+1)p}\partial\xi_{n}^{pp}-k_{p}\mu_{p+1}\xi_{n}^{pp}\partial\psi_{n}^{(p+1)p}+g\,\partial\xi_{n}^{pp}\partial\psi_{n}^{(p+1)p}}\tag{73}$$

$$T_{Pp}^{V}=\frac{k_{p+1}\mu_{p}\left(\xi_{n}^{(p+1)p}\partial\psi_{n}^{(p+1)p}-\psi_{n}^{(p+1)p}\partial\xi_{n}^{(p+1)p}\right)}{k_{p+1}\mu_{p}\xi_{n}^{(p+1)p}\partial\psi_{n}^{pp}-k_{p}\mu_{p+1}\psi_{n}^{pp}\partial\xi_{n}^{(p+1)p}+g\,\partial\psi_{n}^{pp}\partial\xi_{n}^{(p+1)p}}\tag{74}$$

In the above formulas, the subscripts F and P represent the outgoing and incoming waves, respectively. The symbols $T_{(P,F)p}^{H}$ and $R_{(P,F)p}^{H}$ express the transmission and reflection of TE waves (due to the presence of superscript H), whereas the expression $T_{(P,F)p}^{V}$ and $R_{(P,F)p}^{V}$ express the transmission and reflection of the TM waves (due to the presence of superscript V). Using the matrix form:

$$\begin{bmatrix} A_{H,V}^{(p+1)q}+\delta_{p+1}^{q} & B_{H,V}^{(p+1)q} \\ C_{H,V}^{(p+1)q} & D_{H,V}^{(p+1)q} \end{bmatrix}=\begin{bmatrix} \dfrac{1}{T_{Fp}^{H,V}} & \dfrac{R_{Fp}^{H,V}}{T_{Fp}^{H,V}} \\ \dfrac{R_{Pp}^{H,V}}{T_{Pp}^{H,V}} & \dfrac{1}{T_{Pp}^{H,V}} \end{bmatrix}\begin{bmatrix} A_{H,V}^{pq} & B_{H,V}^{pq} \\ C_{H,V}^{pq} & D_{H,V}^{pq}+\delta_{p}^{q} \end{bmatrix}\tag{75}$$

Recursive relations can use to start through: $A_{H,V}^{Nq}=B_{H,V}^{Nq}=C_{H,V}^{1q}=D_{H,V}^{1q}=0$.

It should be noted that due to the difficulty of constructing multiple concentric graphene shells, one can consider each boundary as a dielectric or perfect electric conductor (PEC) in the optical design. For example, an optical absorber consisting of a metal-dielectric spherical resonator whose outermost layer is coated with graphene is shown to enhance the absorption of the resonator [79]. To design optical structures such as absorber and invisible cloaks using Green's function formulation, it is necessary to transfer the point source to infinity to resemble a plane wave. In this case, considering the expansion of Green's function, it can be observed that the only unknown coefficients of expansion are $B_{H,V}^{pq}$ and $D_{H,V}^{pq}$ coefficients. Using the convolution integral it can be shown that [23]:

$$\overline{\overline{\mathbf{G}}}_{0e}.J+\overline{\overline{\mathbf{G}}}_{es}^{(11)}.J=\overline{\overline{\mathbf{G}}}_{es}^{(21)}.J\tag{76}$$

The source dependency is the same in all of the above Green's functions and can be simplified from both sides of the equation. It is observed that the equation is converted into the equation resulting from the Mie analysis of the spherically layered structures. By directly starting from the Mie-Lorentz theory, the same results can be obtained [80]. Moreover, due to the sub-wavelength nature of the localized graphene plasmons, the final formulas can be simplified using the polynomial approximation of the special functions [81]. These structures can also be used as the building blocks of optical meta-materials [82].

3.3.2 Purcell factor and energy transfer between donor-acceptor emitters

As mentioned earlier, one of the important applications of Green's function is studying the interaction of dipole emitters in the vicinity of nanostructures. For this purpose, a vertical dipole in the vicinity of the graphene-based spherical structure, which was introduced in the previous section, is considered and its Purcell factor is calculated using the Green's function. Assuming that the field point and observation points for the dipole with moment $\boldsymbol{d}_0^{\perp} = d_0\hat{\boldsymbol{r}}$ are in the same location of $r' = \Delta > R_1, \theta' = 0$, and $\phi' = 0$, by calculating the scattered field using the convolution integral and the use of [18]:

$$\frac{\Gamma_{total}^{d0}}{\Gamma_0} = 1 + \frac{6\pi\varepsilon_0}{k_1^3 d_0^2}\mathrm{Im}\left[\boldsymbol{d}_0 . E_{sca}^{\boldsymbol{d}_0}(\boldsymbol{r}')\right] \tag{77}$$

where symbol I_m represents the imaginary part of the complex function. The decay rate can be calculated. For this purpose, using relationships:

$$P_n^m(\cos\theta)|_{\theta\to 0} \simeq \sin^m\theta = \begin{cases} 1 & m = 0 \\ 0 & m \neq 0 \end{cases} \tag{78}$$

$$\frac{dP_n^m(\cos\theta)}{d\theta}|_{\theta\to 0} \simeq m \sin^{(m-1)}\theta = \begin{cases} 1 & m = 1 \\ 0 & m \neq 1 \end{cases} \tag{79}$$

The scattered field can be calculated. Thus [73]:

$$\frac{\Gamma_{total}^{\perp}}{\Gamma_0} = 1 - \frac{3}{2}\sum_n n(n+1)(2n+1)\Re\left[\left(\frac{z_n(k_1\Delta)}{k_1\Delta}\right)^2 B_V^{11}\right] \tag{80}$$

As can be seen, the above equation is in full agreement with [83] which has extracted the decay rate for a core-shell plasmonic sphere, whereas in this case, it is necessary to use the Mie coefficients of graphene-based structure, namely, B_V^{11}. Using the derived formulas, the positions of the dipole can be considered arbitrarily. The transferred energy between the donor-acceptor pairs can be calculated straightforwardly using the Green's function $\mathbf{G}$ as [84]:

$$\frac{\Gamma(\omega)}{\Gamma_0(\omega)} = \frac{|\boldsymbol{d}_A . \mathbf{G}(\boldsymbol{r}_A, \boldsymbol{r}_B, \omega) . \boldsymbol{d}_B|^2}{|\boldsymbol{d}_A . \mathbf{G}_0(\boldsymbol{r}_A, \boldsymbol{r}_B, \omega) . \boldsymbol{d}_B|^2} \tag{81}$$

where the subscript 0 refers to the free-space parameters and $\boldsymbol{d}_A$ and $\boldsymbol{d}_B$ are respectively the dipole moments of the acceptor and donor.

4. Conclusion

In conclusion, dyadic Green's function extraction for planarly, cylindrically, and spherically layered medium based on scattering superposition method is a unified approach to deal with a wide range of electromagnetic problems in the realm of biomedicine. Specifically, the interaction of the human skin, body, and head with the electromagnetic sources with arbitrary distributions can be studied. Moreover, by engineering the constitutive parameters of the layers, a variety of novel devices for medical diagnostics and treatment can be proposed. Plasmonic metals and 2D materials are two main categories of such materials. For the sake of efficient analytical analysis, the impedance boundary condition is satisfied in the case of 2D materials to be used in the design of compact devices.

Author details

Shiva Hayati Raad* and Zahra Atlasbaf*
Department of Electrical and Computer Engineering, Tarbiat Modares University, Tehran, Iran

*Address all correspondence to: shiva.hayati@modares.ac.ir
and atlasbaf@modares.ac.ir

References

[1] J. R. Monds and A. G. McDonald, "Determination of skin temperature distribution and heat flux during simulated fires using Green's functions over finite-length scales," *Applied Thermal Engineering*, vol. 50, no. 1, pp. 593–603, 2013.

[2] J. Schmitt, G. Zhou, E. Walker, and R. Wall, "Multilayer model of photon diffusion in skin," *JOSA A*, vol. 7, no. 11, pp. 2141–2153, 1990.

[3] A. Christ, T. Samaras, A. Klingenböck, and N. Kuster, "Characterization of the electromagnetic near-field absorption in layered biological tissue in the frequency range from 30 MHz to 6000 MHz," *Physics in Medicine Biology*, vol. 51, no. 19, p. 4951, 2006.

[4] N. Chahat, M. Zhadobov, L. Le Coq, S. I. Alekseev, and R. Sauleau, "Characterization of the interactions between a 60-GHz antenna and the human body in an off-body scenario," *IEEE Transactions on Antennas Propagation*, vol. 60, no. 12, pp. 5958–5965, 2012.

[5] S. Caorsi, M. Pastorino, and M. Raffetto, "Analytic SAR computation in a multilayer elliptic cylinder for bioelectromagnetic applications," *Bioelectromagnetics: Journal of the Bioelectromagnetics Society, The Society for Physical Regulation in Biology Medicine, The European Bioelectromagnetics Association*, vol. 20, no. 6, pp. 365–371, 1999.

[6] K. S. Nikita, G. S. Stamatakos, N. K. Uzunoglu, and A. Karafotias, "Analysis of the interaction between a layered spherical human head model and a finite-length dipole," *IEEE Transactions on Microwave Theory Techniques,* vol. 48, no. 11, pp. 2003–2013, 2000.

[7] S. Koulouridis and K. S. Nikita, "Study of the coupling between human head and cellular phone helical antennas," *IEEE Transactions on electromagnetic Compatibility*, vol. 46, no. 1, pp. 62-70, 2004.

[8] J. Kim and Y. Rahmat-Samii, "Implanted antennas inside a human body: Simulations, designs, and characterizations," *IEEE Transactions on microwave theory techniques*, vol. 52, no. 8, pp. 1934-1943, 2004.

[9] F. Liu and S. Crozier, "Electromagnetic fields inside a lossy, multilayered spherical head phantom excited by MRI coils: models and methods," Physics in Medicine Biology, vol. 49, no. 10, p. 1835, 2004.

[10] F. Keshmiri and C. Craeye, "Moment-method analysis of normal-to-body antennas using a Green's function approach," *IEEE transactions on antennas propagation*, vol. 60, no. 9, pp. 4259–4270, 2012.

[11] J. S. Meiguni, M. Kamyab, and A. Hosseinbeig, "Theory and experiment of spherical aperture-coupled antennas," *IEEE Transactions on Antennas and Propagation,* vol. 61, no. 5, pp. 2397–2403, 2013.

[12] H. Khodabakhshi and A. Cheldavi, "Irradiation of a six-layered spherical model of human head in the near field of a half-wave dipole antenna," *IEEE transactions on microwave theory techniques*, vol. 58, no. 3, pp. 680-690, 2010.

[13] W. C. Chew, *Waves and fields in inhomogeneous media*. IEEE press, 1995.

[14] J. Sun, C.-F. Wang, J. L.-W. Li, and M.-S. Leong, "Mixed potential spatial domain Green's functions in fast computational form for cylindrically stratified media," *Progress In Electromagnetics Research,* vol. 45, pp. 181–199, 2004.

[15] T. Itoh, "Spectral domain immitance approach for dispersion characteristics

of generalized printed transmission lines," *IEEE transactions on Microwave Theory and Techniques,* vol. 28, no. 7, pp. 733–736, 1980.

[16] M. Thiel and A. Dreher, "Dyadic Green's function of multilayer cylindrical closed and sector-structures for waveguide, microstrip-antenna, and network analysis," *IEEE transactions on microwave theory and techniques,* vol. 50, no. 11, pp. 2576–2579, 2002.

[17] T. V. B. Giang, M. Thiel, and A. Dreher, "A unified approach to the analysis of radial waveguides, dielectric resonators, and microstrip antennas on spherical multilayer structures," *IEEE transactions on microwave theory and techniques,* vol. 53, no. 1, pp. 404–409, 2005.

[18] L. Novotny and B. Hecht, *Principles of nano-optics*. Cambridge university press, 2012.

[19] C. A. Marocico, X. Zhang, and A. L. Bradley, "A theoretical investigation of the influence of gold nanosphere size on the decay and energy transfer rates and efficiencies of quantum emitters," *The Journal of chemical physics,* vol. 144, no. 2, p. 024108, 2016.

[20] V. Karanikolas, C. A. Marocico, and A. L. Bradley, "Spontaneous emission and energy transfer rates near a coated metallic cylinder," *Physical Review A,* vol. 89, no. 6, p. 063817, 2014.

[21] K. Rustomji, R. Abdeddaim, C. M. de Sterke, B. Kuhlmey, and S. Enoch, "Measurement and simulation of the polarization-dependent Purcell factor in a microwave fishnet metamaterial," *Physical Review B,* vol. 95, no. 3, p. 035156, 2017.

[22] A. Vakil, "Transformation optics using graphene: one-atom-thick optical devices based on graphene," 2012.

[23] C.-T. Tai, *Dyadic Green functions in electromagnetic theory*. Institute of Electrical & Electronics Engineers (IEEE), 1994.

[24] L.-W. Cai, "On the computation of spherical Bessel functions of complex arguments," *Computer Physics Communications,* vol. 182, no. 3, pp. 663–668, 2011.

[25] D. Bhattacharya, B. Ghosh, P. K. Goswami, and K. Sarabandi, "Evaluation of Efficient Green's Functions for Spherically Stratified Media," *IEEE Transactions on Antennas and Propagation,* vol. 66, no. 3, pp. 1613–1618, 2018.

[26] Y. S. Cao, L. J. Jiang, and A. E. Ruehli, "An equivalent circuit model for graphene-based terahertz antenna using the PEEC method," *IEEE Transactions on Antennas and Propagation,* vol. 64, no. 4, pp. 1385–1393, 2016.

[27] A. Y. Nikitin, F. J. Garcia-Vidal, and L. Martin-Moreno, "Analytical expressions for the electromagnetic dyadic Green's function in graphene and thin layers," *IEEE Journal of Selected Topics in Quantum Electronics,* vol. 19, no. 3, pp. 4600611–4600611, 2012.

[28] G. W. Hanson, "Dyadic Green's functions and guided surface waves for a surface conductivity model of graphene," *Journal of Applied Physics,* vol. 103, no. 6, p. 064302, 2008.

[29] G. W. Hanson, "Dyadic Green's functions for an anisotropic, non-local model of biased graphene," *IEEE Transactions on antennas and propagation,* vol. 56, no. 3, pp. 747–757, 2008.

[30] M. Cuevas, "Surface plasmon enhancement of spontaneous emission in graphene waveguides," *Journal of Optics,* vol. 18, no. 10, p. 105003, 2016

[31] T. J. Arruda, R. Bachelard, J. Weiner, and P. W. Courteille, "Tunable Fano resonances in the decay rates of a pointlike emitter near a graphene-coated nanowire," *Physical Review B,* vol. 98, no. 24, p. 245419, 2018.

[32] E. L. Tan and S. Y. Tan, "A unified representation of the dyadic Green's

functions for planar, cylindrical and spherical multilayered biisotropic media," *Progress In Electromagnetics Research,* vol. 20, pp. 75–100, 1998

[33] E. L. Tan and S. Y. Tan, "Coordinate-independent dyadic formulation of the dispersion relation for bianisotropic media," *IEEE Transactions on Antennas Propagation,* vol. 47, no. 12, pp. 1820–1824, 1999.

[34] E. L. Tan and S. Y. Tan, "Concise spectral formalism in the electromagnetics of bianisotropic media," *Progress In Electromagnetics Research,* vol. 25, pp. 309–331, 2000.

[35] E. Tan, "Vector wave function expansions of dyadic Green's functions for bianisotropic media," *IEE Proceedings-Microwaves, Antennas propagation,* vol. 149, no. 1, pp. 57–63, 2002.

[36] W.-B. Lu, H. Chen, and Z.-G. Liu, "A review of microwave devices based on CVD-grown graphene with experimental demonstration," *EPJ Applied Metamaterials*, vol. 6, p. 8, 2019.

[37] J. Zhang, Z. Liu, W. Lu, H. Chen, B. Wu, and Q. Liu, "A low profile tunable microwave absorber based on graphene sandwich structure and high impedance surface," *International Journal of RF Microwave Computer-Aided Engineering,* vol. 30, no. 2, p. e22022, 2020.

[38] D. Yi, X.-C. Wei, and Y.-L. Xu, "Tunable microwave absorber based on patterned graphene," *IEEE Transactions on Microwave Theory Techniques,* vol. 65, no. 8, pp. 2819–2826, 2017.

[39] Z. Fang *et al.*, "Active tunable absorption enhancement with graphene nanodisk arrays," vol. 14, no. 1, pp. 299–304, 2014.

[40] B. Wu *et al.*, "Experimental demonstration of a transparent graphene millimetre wave absorber with 28% fractional bandwidth at 140 GHz," *Scientific reports,* vol. 4, p. 4130, 2014.

[41] C. Srichan *et al.*, "Highly-sensitive surface-enhanced Raman spectroscopy (SERS)-based chemical sensor using 3D graphene foam decorated with silver nanoparticles as SERS substrate," *Scientific reports,* vol. 6, p. 23733, 2016.

[42] J. B. Maurya and Y. K. Prajapati, "Experimental Demonstration of DNA Hybridization Using Graphene-Based Plasmonic Sensor Chip," *Journal of Lightwave Technology*, 2020.

[43] L. A. Falkovsky, "Optical properties of graphene and IV–VI semiconductors," *Physics-Uspekhi,* vol. 51, no. 9, p. 887, 2008.

[44] H. Nasari and M. Abrishamian, "Nonlinear terahertz frequency conversion via graphene microribbon array," *Nanotechnology,* vol. 27, no. 30, p. 305202, 2016.

[45] S. Thongrattanasiri, F. H. Koppens, and F. J. G. De Abajo, "Complete optical absorption in periodically patterned graphene," *Physical review letters,* vol. 108, no. 4, p. 047401, 2012.

[46] S. H. Raad, Z. Atlasbaf, M. Shahabadi, and J. Rashed-Mohassel, "Dyadic Green's Function for the Tensor Surface Conductivity Boundary Condition," *IEEE Transactions on Magnetics,* vol. 55, no. 11, pp. 1–7, 2019.

[47] Y. R. Padooru, A. B. Yakovlev, C. S. Kaipa, G. W. Hanson, F. Medina, and F. Mesa, "Dual capacitive-inductive nature of periodic graphene patches: Transmission characteristics at low-terahertz frequencies," *Physical Review B,* vol. 87, no. 11, p. 115401, 2013.

[48] A. Andryieuski and A. V. Lavrinenko, "Graphene metamaterials based tunable terahertz absorber: effective surface conductivity approach," *Optics express,* vol. 21, no. 7, pp. 9144–9155, 2013.

[49] Y. Guo, T. Zhang, W.-Y. Yin, and X.-H. Wang, "Improved hybrid FDTD method for studying tunable graphene frequency-selective surfaces (GFSS) for THz-wave applications," *IEEE Transactions on Terahertz Science and Technology,* vol. 5, no. 3, pp. 358–367, 2015.

[50] J. S. Gomez-Diaz, M. Tymchenko, and A. Alu, "Hyperbolic plasmons and topological transitions over uniaxial metasurfaces," *Physical Review Letters,* vol. 114, no. 23, p. 233901, 2015.

[51] J. Gomez-Diaz and A. Alu, "Flatland optics with hyperbolic metasurfaces," *ACS Photonics,* vol. 3, no. 12, pp. 2211–2224, 2016.

[52] A. Fallahi and J. Perruisseau-Carrier, "Design of tunable biperiodic graphene metasurfaces," *Physical Review B,* vol. 86, no. 19, p. 195408, 2012.

[53] D. L. Sounas and C. Caloz, "Gyrotropy and nonreciprocity of graphene for microwave applications," *IEEE Transactions on Microwave Theory and Techniques,* vol. 60, no. 4, pp. 901–914, 2012.

[54] L. Lin, L.-S. Wu, W.-Y. Yin, and J.-F. Mao, "Modeling of magnetically biased graphene patch frequency selective surface (FSS)," in *2015 IEEE MTT-S International Microwave Workshop Series on Advanced Materials and Processes for RF and THz Applications (IMWS-AMP)*, 2015, pp. 1–3: IEEE.

[55] Y. Yermakov *et al.*, "Effective surface conductivity of optical hyperbolic metasurfaces: from far-field characterization to surface wave analysis," *Scientific reports,* vol. 8, no. 1, p. 14135, 2018.

[56] J. Komijani, J. Rashed-Mohassel, and A. Mirkamali, "Dyadic Green Functions for a Dielectric Layer on a PEMC Plane," *Progress In Electromagnetics Research,* vol. 6, pp. 9–22, 2009.

[57] J. Gómez-Díaz and J. Perruisseau-Carrier, "Propagation of hybrid transverse magnetic-transverse electric plasmons on magnetically biased graphene sheets," *Journal of Applied Physics,* vol. 112, no. 12, p. 124906, 2012.

[58] Raad, Shiva Hayati, Zahra Atlasbaf, and Mauro Cuevas. "Dyadic Green's function for the graphene-dielectric stack with arbitrary field and source points." arXiv preprint arXiv: 2104.03581 (2021).

[59] M. Cuevas, M. A. Riso, and R. A. Depine, "Complex frequencies and field distributions of localized surface plasmon modes in graphene-coated subwavelength wires," *Journal of Quantitative Spectroscopy and Radiative Transfer,* vol. 173, pp. 26–33, 2016.

[60] M. Hajati and Y. Hajati, "Plasmonic characteristics of two vertically coupled graphene-coated nanowires integrated with substrate," *Applied Optics*, vol. 56, no. 4, pp. 870–875, 2017.

[61] S. H. Raad and Z. Atlasbaf, "Dyadic analysis of a cylindrical wire consisting of a cover with fully-populated surface conductivity tensor," *Optics express,* vol. 27, no. 15, pp. 21214–21225, 2019.

[62] J. Gomez-Diaz, M. Tymchenko, and A. Alù, "Hyperbolic metasurfaces: surface plasmons, light-matter interactions, and physical implementation using graphene strips," *Optical Materials Express,* vol. 5, no. 10, pp. 2313–2329, 2015.

[63] K.-H. Kim, Y.-S. No, S. Chang, J.-H. Choi, and H.-G. Park, "Invisible hyperbolic metamaterial nanotube at visible frequency," *Scientific reports,* vol. 5, p. 16027, 2015.

[64] Y. Mazor and A. Alù, "Angular-momentum selectivity and asymmetry in highly confined wave propagation along sheath-helical metasurface tubes," *Physical Review B,* vol. 99, no. 15, p. 155425, 2019.

[65] Y. Huang, Y. Shen, C. Min, and G. Veronis, "Switching photonic nanostructures between cloaking and superscattering regimes using phase-change materials," *Optical Materials Express,* vol. 8, no. 6, pp. 1672–1685, 2018.

[66] Y. Hua and T. K. Sarkar, "Generalized pencil-of-function method for extracting poles of an EM system from its transient response," 1989.

[67] C. Tokgoz and G. Dural, "Closed-form Green's functions for cylindrically stratified media," *IEEE Transactions on Microwave theory and Techniques,* vol. 48, no. 1, pp. 40–49, 2000.

[68] S. H. Raad, C. J. Zapata-Rodríguez, and Z. Atlasbaf, "Multi-frequency super-scattering from sub-wavelength graphene-coated nanotubes," *JOSA B,* vol. 36, no. 8, pp. 2292–2298, 2019.

[69] S. H. Raad, C. J. Zapata-Rodríguez, and Z. Atlasbaf, "Graphene-coated resonators with frequency-selective super-scattering and super-cloaking," *Journal of Physics D: Applied Physics,* vol. 52, no. 49, p. 495101, 2019.

[70] S. H. Raad, Z. Atlasbaf, and M. Cuevas, "Scattering from Multilayered Graphene-Based Cylindrical and Spherical Particles," in *Nanoplasmonics*: IntechOpen, 2020.

[71] X. Ma *et al.*, "Graphene oxide wrapped gold nanoparticles for intracellular Raman imaging and drug delivery," vol. 1, no. 47, pp. 6495–6500, 2013.

[72] D. Cai, L. Ding, S. Wang, Z. Li, M. Zhu, and H. Wang, "Facile synthesis of ultrathin-shell graphene hollow spheres for high-performance lithium-ion batteries," *Electrochimica Acta*, vol. 139, pp. 96-103, 2014.

[73] S. H. Raad, Z. Atlasbaf, C. J. Zapata-Rodríguez, M. Shahabadi, and J. Rashed-Mohassel, "Dyadic Green's function for the electrically biased graphene-based multilayered spherical structures," *Journal of Quantitative Spectroscopy Radiative Transfer,* vol. 256, p. 107251, 2020.

[74] Z.-S. Wu, Z.-J. Li, H. Li, Q.-K. Yuan, and H.-Y. Li, "Off-axis Gaussian beam scattering by an anisotropic coated sphere," *IEEE Transactions on Antennas and Propagation,* vol. 59, no. 12, pp. 4740–4748, 2011.

[75] L.-W. Li, P.-S. Kooi, M.-S. Leong, and T.-S. Yee, "Electromagnetic dyadic Green's function in spherically multilayered media," *IEEE Transactions on Microwave Theory and Techniques,* vol. 42, no. 12, pp. 2302–2310, 1994.

[76] M. Farhat, C. Rockstuhl, and H. Bağcı, "A 3D tunable and multi-frequency graphene plasmonic cloak," *Optics express,* vol. 21, no. 10, pp. 12592–12603, 2013.

[77] F. TING, "Fast solutions of electromagnetic fields in layered media," 2008.

[78] J. D. Shumpert, "Modeling of Periodic Dielectric Structures (electomagnetic Crystals)," University of Michigan, 2001.

[79] M. Wan *et al.*, "Strong tunable absorption enhancement in graphene using dielectric-metal core-shell resonators," *Scientific reports,* vol. 7, no. 1, p. 32, 2017.

[80] S. H. Raad, Z. Atlasbaf, J. Rashed-Mohassel, and M. Shahabadi, "Scattering from graphene-based multilayered spherical structures," *IEEE Transactions on Nanotechnology,* vol. 18, pp. 1129–1136, 2019.

[81] S. H. Raad and Z. Atlasbaf, "Equivalent RLC ladder circuit for scattering by graphene-coated nanospheres," *IEEE Transactions on Nanotechnology,* vol. 18, pp. 212–219, 2019.

[82] S. H. Raad and Z. Atlasbaf, "Tunable optical meta-surface using graphene-

coated spherical nanoparticles," *AIP Advances,* vol. 9, no. 7, p. 075224, 2019.

[83] T. J. Arruda, A. S. Martinez, F. A. Pinheiro, R. Bachelard, S. Slama, and P. W. Courteille, "Fano resonances in plasmonic core-shell particles and the Purcell effect," in *Fano Resonances in Optics and Microwaves*: Springer, 2018, pp. 445–472.

[84] V. D. Karanikolas and A. L. Bradley, "Interactions Between Quantum Emitters in the Presence of Plasmonic Nanostructures," Trinity College Dublin, 2016.

Chapter 2

A TLM Formulation Based on Fractional Derivatives for Dispersive Cole-Cole Media

Mohammed Kanjaa, Otman El Mrabet and Mohsine Khalladi

Abstract

An auxiliary differential equation (ADE) transmission line method (TLM) is proposed for broadband modeling of electromagnetic (EM) wave propagation in biological tissues with the Cole-Cole dispersion Model. The fractional derivative problem is surmounted by assuming a linear behavior of the polarization current when the time discretization is short enough. The polarization current density is approached using Lagrange extrapolation polynomial and the fractional derivation is obtained according to Riemann definition of a fractional α-order derivative. Reflection coefficients at an air/muscle and air/fat tissues interfaces simulated in a 1-D domain are found to be in good agreement with those obtained from the analytic model over a broad frequency range, demonstrating the validity of the proposed approach.

Keywords: computational electromagnetics, numerical methods, transmission line matrix, biological system modeling, Cole-Cole medium, fractional derivative

1. Introduction

In the last two decades there has been a growing interest in the interaction between biological tissues and electromagnetic field at microwave frequencies. New promising applications of this technology in biomedical engineering like microwave imaging [1, 2], minimal invasive cancer therapies as thermal ablations [3], ultra-wide band temperature dependent dielectric spectroscopy [4] and EM dosimetry [5], rely heavily on an accurate mathematical model of the response of these tissues to an external electromagnetic field. Numerical resolution of the propagation problem within these tissues requires a robust mathematical model and the previous incorporation of the dielectric data that at all the working frequencies.

A first model of the time response in a time varying electric field of biological tissues was formulated by Debye [6] through a time decaying polarization current $j(t) = \frac{\partial P}{\partial t}$ resulting from the time variation of the polarization vector $P(t)$, the permittivity for this model is given in Eq. (1) and the corresponding argand diagram is depicted in **Figure 1**.

$$\varepsilon_r^*(\omega) - \varepsilon_{r\infty} = \varepsilon_r(\omega) - \varepsilon_{r\infty} - j\varepsilon_r'(\omega) = \frac{\Delta\varepsilon_p}{1 + j\omega\tau_p} \tag{1}$$

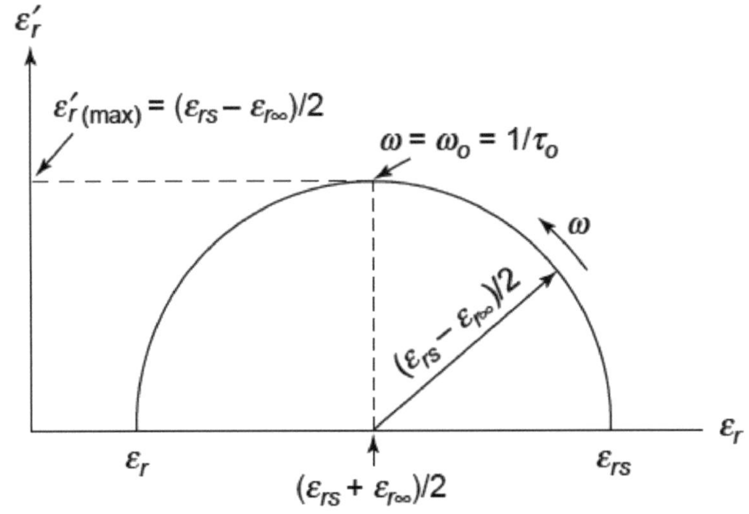

Figure 1.
Argand diagram for a unique pole Debye model obtained by representing Eq. (1) in the complexe plane.

This model even if it fits the experimental results in liquids it loses its accuracy when applied over a large band of frequency or in the presence of more than one type of polar molecule. This non-Debye relaxation is attributed to the existence of different relaxation processes [7] each with its own relaxation time τ and its amplitude $\Delta\varepsilon$. A more accurate approximation to the behavior of this category of dielectrics is given by the sum of the individual relaxation processes leading to a multipole model, the permittivity is given as the sum of the p terms corresponding to the p poles each with its own amplitude. In the case of the biological tissue there is a multiple contribution of each relaxation process resulting in a broadening of the relaxation zone. Cole-Cole [8] proposed an Argand diagram in which in which ε_r the imaginary part of the complex permittivity is plotted as a function of its real part ε_r, following the empirical relation:

$$\varepsilon_r^*(\omega) - \varepsilon_{r\infty} = \varepsilon_r(\omega) - \varepsilon_{r\infty} - j\varepsilon_r'(\omega) = \sum_{p=1}^{p} \frac{\Delta\varepsilon_p}{1 + (j\omega\tau_p)^{\alpha_p}} \quad (2)$$

where $\varepsilon_{r\infty}$ is the optical relative permittivity, $\Delta\varepsilon_p = \varepsilon_{rs} - \varepsilon_{r\infty}$ is the amplitude of the $p - th$ pole, ε_s is the static relative permittivity, ω is the angular frequency , τ_p is the relaxation time, α_p the parameter that indicates the broadening of the dispersion for this pole, which in the Cole-Cole model must satisfy $0 < \alpha_p < 1$, in the case of $\alpha_p = 1$ the model is simplified to a Debye dispersion problem and $j = \sqrt{-1}$ (**Figure 2**).

The difficulty in the numerical implementation of such a model arises from the α_p parameter which is a noninteger. The consequence is a fractional order differentiation in the time domain which is more challenging compared to the Debye time domain solution. To address this problem, authors in [9] used the Letnikov fractional derivative by introducing a Stirling asymptotic formula and a recursive relation for the polarization vector. The computational demands of the FDTD scheme were considerably reduced, but nevertheless it required the storage of a large number of previous values of the polarization vector. In [10], authors used the Z transform to formulate the frequency dependence between the electric flux density and the electric field, the fractional derivative was approximated by using a polynomial method. Rekanos et al. [11] used the Pade approximation, where the

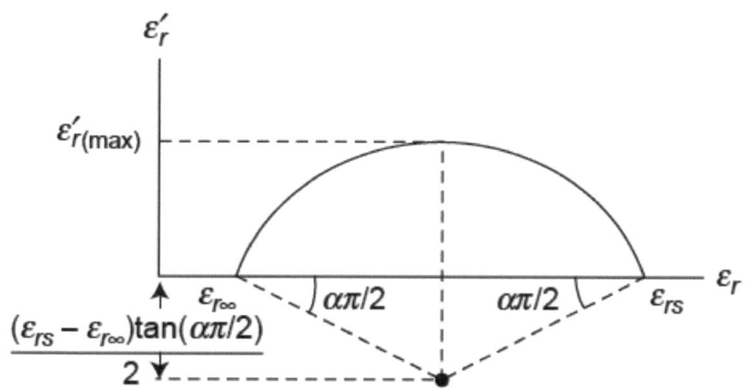

Figure 2.
Argand diagram for a simple pole Cole-Cole model obtained by representing Eq. (2) in the complexe plane.

fractional power term is approximated using a Pade technique. All the latter approximations were made in the context of the FDTD method.

The TLM method, even if it is flexible and wide band and being a time domain method, cannot deal with the dispersive aspect of the Cole-Cole medium directly. In [12] an approach based on a convolution product between the susceptibility and the electric field and the temporal behavior is deduced by a DFT and a nonrecursive summation leading to a considerable computational cost and a nonnegligible error at high frequencies. The causality principle is used to justify the minor dependence of the recent susceptibility values on previous ones, but the problem of the fractional Differentegration wasn't addressed.

In this work an ADE-TLM algorithm to model the Cole-Cole dispersion. The polarization current density is approached using an extrapolation with Lagrange polynomial method and the fractional derivation is obtained using the Riemann definition of the α-order derivative. The auxiliary differential equation is used to establish the update equation of the polarization currents which are included later in the general structure of the SCN-TLM node.

2. Formulation of the method

In the Cole-Cole model the relationship between the polarization current related to the p_{th} pole and the electric field is given by:

$$\mathbf{J}_p(\omega) = \varepsilon_0 \Delta\varepsilon_p \frac{j\omega}{1 + \left(j\omega\tau_p\right)^{\alpha_p}} \mathbf{E}(\omega) \tag{3}$$

where $\left(j\omega\tau_p\right)^{\alpha_p}$ is the power law function of frequency which in time domain results in a fractional derivative of order α:

$$\mathbf{J}_p(t) + \tau^{\alpha_p} D^{\alpha_p} \mathbf{J}_p(t) = \varepsilon_0 \Delta\varepsilon_p \frac{\partial \mathbf{E}(t)}{\partial t} \tag{4}$$

One could consider an analytic solution for this equation, but due to the fractional derivative this can only be done for particular values of α_p, hence a numerical solution is necessary with a previous discretization of the temporal values of the vectors.

In order to obtain the discretized expression of $\mathbf{J}_p(t)$, the polarization current vector can be approached using a Lagrange interpolation in the time interval $t_i < t < t_j$ with cardinal functions:

$$L_i(t) = \prod_{j=1;i\neq j}^{n} \frac{(t-t_j)}{(t_i-t_j)} \tag{5}$$

the interpolated expression of $\mathbf{J}_p(t)$ is then obtained:

$$\mathbf{J}_p(t) = \sum_{i=1}^{n} \mathbf{J}^i L_i(t) \tag{6}$$

and by applying (5) in the time interval $n\Delta t < t < (n+1)\Delta t$ we obtain the cardinal functions:

$$L_n(t) = \frac{t-(n+1)\Delta t}{n\Delta t-(n+1)\Delta t} \qquad L_{n+1}(t) = \frac{t-n\Delta t}{(n+1)\Delta t-n\Delta t} \tag{7}$$

By substituting (7) in (6) the polynomial extrapolation of the polarization current density between time steps $n\Delta t$ and $(n+1)\Delta t$ can be found in a straightforward manner as follow:

$$\mathbf{J}_p(t) = \frac{t-(n+1)\Delta t}{n\Delta t-(n+1)\Delta t}\mathbf{J}_p^n + \frac{t-n\Delta t}{(n+1)\Delta t-n\Delta t}\mathbf{J}_p^{n+1} \tag{8}$$

which after simplification can be rewritten as:

$$\mathbf{J}_p(t) = \frac{\mathbf{J}_p^{n+1}-\mathbf{J}_p^n}{\Delta t}t + (n+1)\mathbf{J}_p^n - n\mathbf{J}_p^{n+1} \tag{9}$$

Furthermore, by substituting (8) into (3) we derive the auxiliary differential equation with a fractional order derivative given by:

$$\frac{\mathbf{J}_p^{n+1}+\mathbf{J}_p^n}{2} + \tau^{\alpha_p}\frac{\mathbf{J}_p^{n+1}-\mathbf{J}_p^n}{\Delta t}D^{\alpha_p}t + D^{\alpha_p}\left((n+1)\mathbf{J}_p^n - n\mathbf{J}_p^{n+1}\right) = \varepsilon_0\Delta\varepsilon_p\frac{\partial\mathbf{E}}{\partial t} \tag{10}$$

The generalization of the derivation operator to arbitrary non-integer orders, has been subject to intensive research from mathematicians [13, 14] and in electromagnetism [15]. One of the definitions in [16] for the fractional derivative, symbolized by the operator $\mathbf{D}_x^\alpha$ is given for the power series with fractional exponents by considering a function $h(x)$ defined as a power series $h(x) = \sum_{i=1}^{n} A_i(x-a)^{\nu+i/n}$ for $\nu > -1$ and n a positive integer, so each term $(x-a)^p$ could have a noninteger exponent $p = \nu + i/n$, then according to Reimann [15] a fractional derivative of order α for this term is given by:

$$D_x^\alpha(x-a)^p = \frac{d^\alpha(x-a)^p}{d(x-a)^\alpha} = \frac{\Gamma(p+1)}{\Gamma(p-\alpha+1)}(x-a)^{p-\alpha} \quad for \quad x > a \tag{11}$$

where in our case α is a noninteger number [16].

Hence by applying the fractional derivation operator given in (11) to (10) and for $p = 1$ and $a = 0$ we obtain the update equation for the polarization current:

$$\mathbf{J}_p^{n+1} = -\frac{\phi_p}{\psi_p}\mathbf{J}_p^n + \frac{\zeta_p}{\psi_p}\left(\mathbf{E}^{n+1} - \mathbf{E}^n\right) \tag{12}$$

where the update parameters ψ_p, ϕ_p and ζ_p are:

$$\psi_p = \frac{1}{2} + \frac{\tau_p^{\alpha_p}}{\Delta t}\frac{\Gamma(2)}{\Gamma(2-\alpha_p)}t^{1-\alpha_p} - n\frac{\tau_p^{\alpha_p}}{\Gamma(1-\alpha_p)}t^{-\alpha_p} \tag{13}$$

$$\phi_p = \frac{1}{2} - \frac{\tau_p^{\alpha_p}}{\Delta t}\frac{\Gamma(2)}{\Gamma(2-\alpha_p)}t^{1-\alpha_p} + (n+1)\frac{\tau_p^{\alpha_p}}{\Gamma(1-\alpha_p)}t^{-\alpha_p} \tag{14}$$

$$\zeta_p = \frac{\varepsilon_0 \Delta\varepsilon_p}{\Delta t} \tag{15}$$

To get the update equation of the electric field components we start from the Maxwell-Ampere equation, and by including the conductivity term:

$$\nabla \times \mathbf{H} = \varepsilon_0 \varepsilon_\infty \frac{\partial \mathbf{E}}{\partial t} + \sigma_0 \mathbf{E} + \sum_{p=1}^{P} \mathbf{J}_p \tag{16}$$

where σ_0 is the static ionic conductivity, Eq. (16) formulated at time step $n + \frac{1}{2}$ and by approaching $\mathbf{J}^{n+\frac{1}{2}}$ by the average of its values at time steps n and $n + 1$

$$\mathbf{J}_p^{n+\frac{1}{2}} = \frac{\mathbf{J}_p^{n+1} + \mathbf{J}_p^n}{2} \tag{17}$$

Finally, the updated electric field is given by:

$$E_{x,y,z}^{n+1} = E_{x,y,z}^n\left(1 - \frac{2\sigma_0 \Delta t}{D}\right) - \sum_p \frac{\left(1 - \frac{\phi_p}{\psi_p}\right)\Delta t}{D} J_{px,y,z}^n + \frac{2\Delta t}{D}(\nabla \times \mathbf{H})_{x,y,z}^{n+\frac{1}{2}} \tag{18}$$

$$D = 2\varepsilon_0 \varepsilon_\infty + \sum_p \frac{\zeta_p \Delta t}{\psi_p} + \sigma_0 \Delta t \tag{19}$$

3. The TLM Formalisme

The TLM method is a numerical technique based on the discretization of the computational domain according to Huygens principle as an alternative to the Maxwell equations used in the FDTD method [17]. In this method the simulation domain is discretized in cells where a series of uniform transmission lines, parallel and series stubs and additional sources are used to take account of the real characteristics of the propagation through the medium as given by Maxwell equations. Therefore, instead of electric and magnetic field components the electromagnetic field is represented by voltage and current waves, propagating through the unit cell circuit referred to as symmetrical condensed node (SCN). The relationship between the electromagnetic field and the voltage and current waves at the time step $n\Delta t$ and at the center of the node are formulated as follows [18].

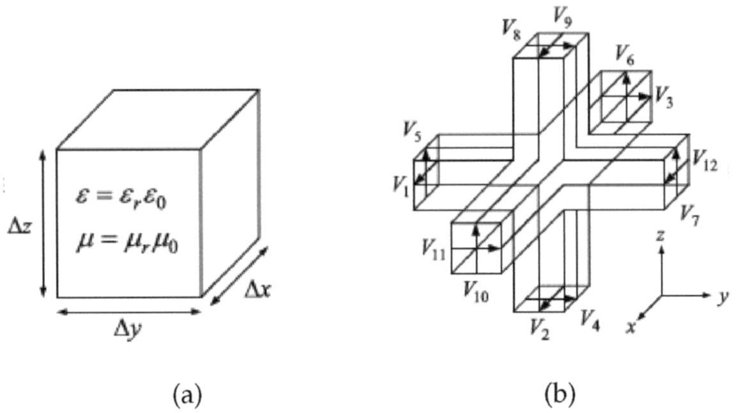

Figure 3.
Structure of the TLM symmetrical condensed node, (a) unit volume of thedielectric material with dimensions Δx, Δy,and Δz, (b) equivalent SCN.

As can be seen in **Figure 3(b)**, the SCN consists of interconnected transmission lines. This structure models a unit cell of the propagation medium as in **Figure 3(a)**. Each face of the cell corresponds to two ports orthogonal to each other and labeled from 1 to 12, these stubs model the propagation through free space and have therefore its characteristic impedance $Z_0 = \sqrt{\varepsilon_0 \mu_0}$. Three additional open circuit (13,14,15) stubs must be added to the node to account for the dispersive behavior of the medium. The SCN nodes are connected to each other to form the simulation domain given in **Figure 4** for a 1-D propagation.

The relationship between the electromagnetic field and the voltage and current waves at the time step $n\Delta t$ and at the center of the node are formulated as follows [18]:

$$E^n_{x,y,z} = \frac{V^n_{x,y,z}}{\Delta l} \qquad H^n_{x,y,z} = \frac{I^n_{x,y,z}}{\Delta l} \tag{20}$$

In this algorithm as in [11, 19] the dispersive properties of the medium are accounted for by adding voltage sources ${}_sV$ to each node, these sources constitute the counterparts of the terms in the FDTD formulation that have no equivalents in the TLM formalism. Therefore the voltage at the center of the node, as deduced from the conservation laws [20], becomes:

$$V^{n+1}_{x,y,z} = V^n_{x,y,z} + \frac{1}{4 + Y_{ocx,y,z}}\left({}_sV^{n+1}_{x,y,z} + {}_sV^n_{x,y,z}\right) + \frac{4}{4 + Y_{ocx,y,z}} \frac{\Delta l \Delta t}{\varepsilon_0} (\nabla \times \mathbf{H})^{n+\frac{1}{2}}_{x,y,z} \tag{21}$$

where $Y_{ocx,y,z}$ indicates the normalized admittance of the open circuit stub added to the node.

When expressed in terms of incident voltages on the corresponding stubs and accounting for the voltage sources injected in the stubs (16,17,18) to complete the model of the Cole-Cole medium:

$$V^{n+1}_{x,y,z} = \frac{2}{4 + Y_{ox,y,z}}\left[V^i_{1,3,5} + V^i_{2,4,6} + V^i_{9,8,7} + V^i_{12,11,10} + Y_{oc,x,y,z} V^i_{13,14,15} + \frac{1}{2}{}_sV^{n+1}_{16,17,18}\right] \tag{22}$$

where the subscript i indicates the incident pulses on the indicated stubs.

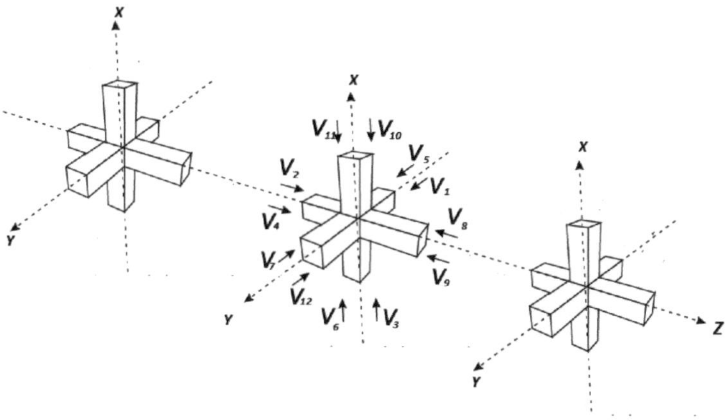

Figure 4.
Structure of the 1-D SCN grid used in this work to model the simulation domain.

In the TLM formalism the update equation issued from the ADE in the Eq. (12) becomes:

$$J^{n+1}_{px,y,z} = -\frac{\phi_p}{\psi_p} J^n_{px,y,z} + \frac{\zeta_p}{\psi_p \Delta l}\left(V^{n+1}_{x,y,z} - V^n_{x,y,z}\right) \tag{23}$$

and the electric field update equation is also formulated in the TLM formalism as:

$$V^{n+1}_{x,y,z} = V^n_{x,y,z}\left(1 - \frac{2\sigma_0 \Delta t}{D}\right) - \sum_p \frac{\left(1 - \frac{\phi_p}{\psi_p}\right)\Delta t \Delta l}{D} J^n_{px,y,z} + \frac{2\Delta t \Delta l}{D}(\nabla \times \mathbf{H})^{n+\frac{1}{2}}_{x,y,z} \tag{24}$$

The analogy between Eqs. (22) and (24) the expression of the normalized admittance of the stub added to the SCN node is obtained straightforward:

$$Y_{oc} = 4\left(\frac{D}{2\varepsilon_0} - 1\right) \tag{25}$$

and the update equation for voltage sources at the center of the node:

$${}_sV^{n+1}_{x,y,z} = -{}_sV^n_{x,y,z} - \frac{4\sigma_0 \Delta t}{\varepsilon_0} V^n_{x,y,z} - \sum_p \frac{2\left(1 - \frac{\phi_p}{\psi_p}\right)\Delta t \Delta l}{\varepsilon_0} J_{px,y,z} \tag{26}$$

that can be simplified to

$${}_sV^{n+1}_{x,y,z} = -{}_sV^n_{x,y,z} + C_1 V^n_{x,y,z} + \sum_p C_{2p} J_{px,y,z} \tag{27}$$

with update constants:

$$C_1 = -\frac{4\sigma_0 \Delta t}{\varepsilon_0} \tag{28}$$

$$C_{2p} = -\frac{2\left(1 - \frac{\phi_p}{\psi_p}\right)\Delta t \Delta l}{\varepsilon_0} \tag{29}$$

A final step to establish the TLM formulation is to express the scattering process on the capacitive stubs:

$$\left[V^r_{13,14,15}\right]^n = \left[V_{x,y,z}\right]^n - \left[V^i_{13,14,15}\right]^n \tag{30}$$

4. Simulation and results

In order to verify the new proposed algorithm, and for the sake of simplicity a one dimensional problem is simulated, where a Cole-Cole medium (fat or muscle) occupies the region $z \geq 0$, while the rest of space is air, 2 dimensional and 3 dimensional propagation and reflection models can be naturally deducted by ensuring the connexion between TLM cells in the y axis using ports 1,5,7 and 12, to account for the propagation along the x axis ports 3,6,10 and 11 must be connected to adjacent cells as in **Figure 5**.

The incident wave is a derivative Gaussian pulse given by $\vec{E}^{inc} = \frac{(t-t_0)}{\tau^2} exp\left(-4\pi\left(\frac{(t-t_0)^2}{\tau^2}\right)\hat{e}_x\right.$ where $t_0 = 200\Delta t$ and $\tau = 220\Delta t$ polarized in the x direction, and propagates along the z axis. In the TLM implementation, the considered 1-D simulation domain (see **Figure 6**) consists of a grid of 1000 cells interconnected on the z axis as in **Figure 4**, 500 of which were used to model the Cole-Cole medium

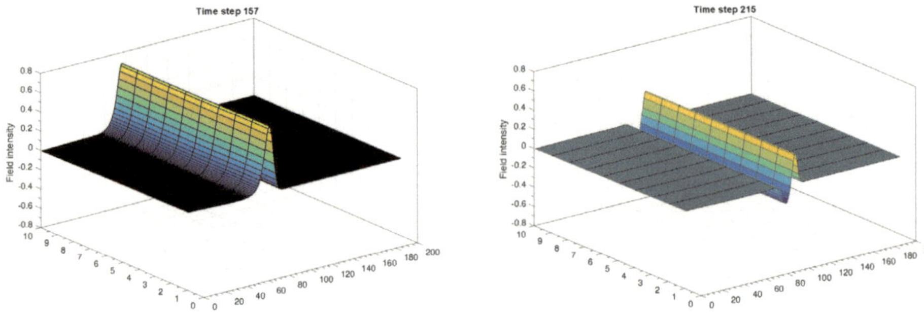

(a) (b)

Figure 5.
2-D ADE-TLM simulation of an incident and reflected pulse on the air/muscle, the tissue is modeled by a 4 pole Cole-Cole model, the interface is located at the TLM cell z = 100, at time steps 160 (a) and 218 (b).

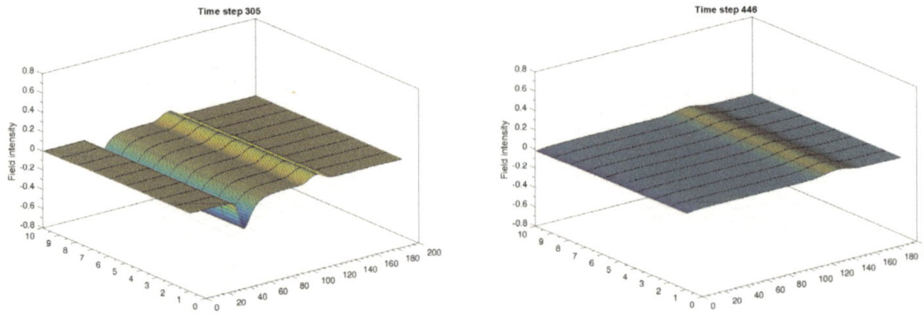

(a) (b)

Figure 6.
2-D ADE-TLM simulation of a reflected and transmitted pulse on the air/muscle , the tissue is modeled by a 4 pole Cole-Cole model, the interface is located at the TLM cell z = 100 , at (a) time step 305, and (b) at time step 446.

and the remaining cells were used to model the air. The cell dimension Δl satisfies the stability condition $\Delta t = \Delta l/2c$, the time step is set as $\Delta t = 0.125ps$. The simulation domain is truncated by an unsplit PML layer [21] of 10 cells at the beginning and the end.

The fourth order Cole-Cole parameters for fat tissue as well as for muscle tissue are listed in **Table 1** [9]. The values of the electric field are recorded at a point P located at the center of the SCN node 10 cells before the air-medium interface. **Figure 7(a)** shows the incident impulse at iteration 399 propagating in the air and **Figure 7(b)** at iteration 927 depicts the reflected and transmitted impulse on the air/muscle interface.

Tissue	ε_∞	σ_0	$\Delta\varepsilon_1$	τ_1 (ps)	α_1	$\Delta\varepsilon_2$	τ_2 (ps)	α_2	$\Delta\varepsilon_3$	τ_3 (μs)	α_3	$\Delta\varepsilon_4$	τ_4 (ps)	$\Delta\varepsilon_4$
Muscle	2.5	0.035	9	7.96	0.8	35	15.92	0.9	3.3E4	159.15	0.95	1.0E7	15.915	0.99
Fat	4	0.2	50	7.23	0.9	7000	353.68	0.9	1.2E6	318.31	0.9	2.5E7	2.274	1

Table 1.
Cole-Cole constants for tissues of muscle and fat.

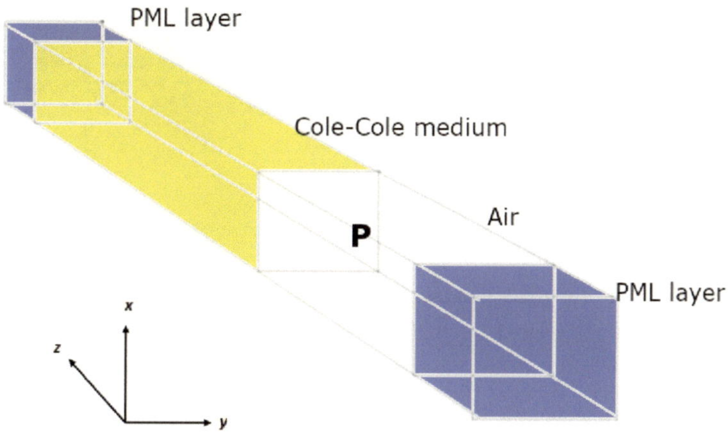

Figure 7.
Propagation of an incident derivative Gaussian pulse through Cole-Cole medium model of human muscle (a) at iteration 399 (incident pulse) (b) at iteration 677 (reflected and transmitted pulse).

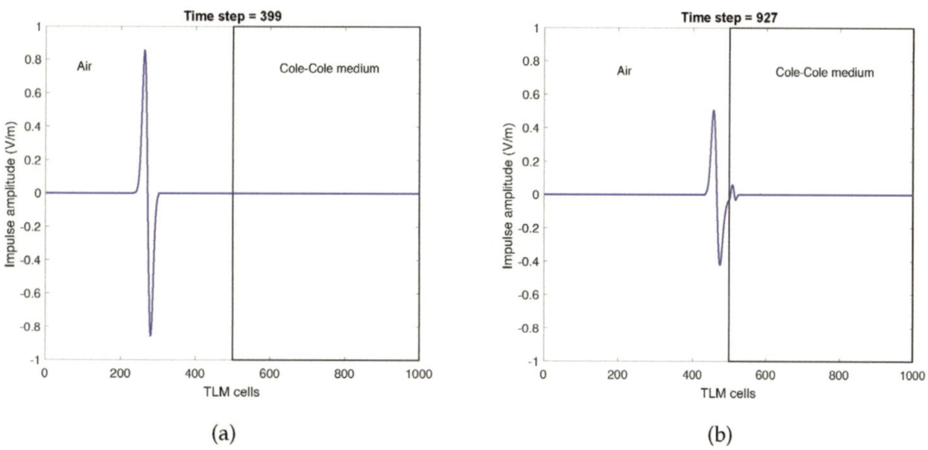

Figure 8.
Reflection coefficient (dB) at normal incidence on an air/muscle interface.

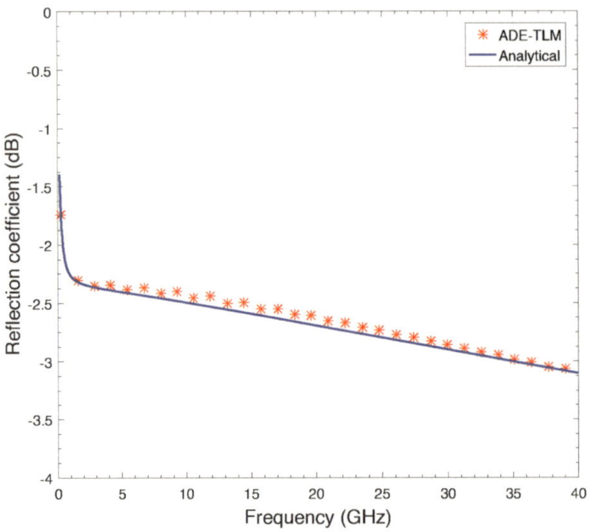

Figure 9.
Reflection coefficient (dB) at normal incidence on an air/fat interface.

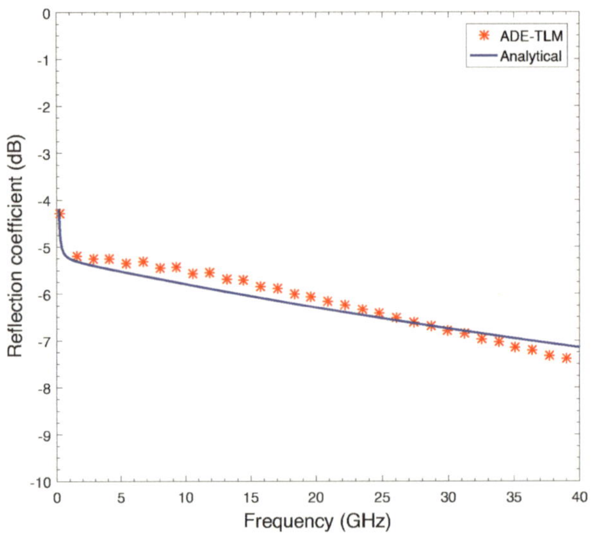

Figure 10.
2-D ADE-TLM simulation of a reflected and transmitted pulse on the air/muscle, the tissue is modeled by a 4 pole Cole-Cole model, the interface is located at the TLM cell $z = 100$, at time steps are 305 and 446.

In **Figures 8** and **9**, the simulation results for both cases (fat and muscle) are compared to the theoretical reflection at the air/Cole-Cole medium interface which can be obtained by using the following equation [22]:

$$|R| = \frac{|1 - \sqrt{\varepsilon_r}|}{|1 + \sqrt{\varepsilon_r}|} \tag{31}$$

A good agreement over the whole frequency band is observed. The slight discrepancy between the numerical and theoretical results can be ascribed to the approximation made to the polarization current value when performing the Lagrange extrapolation on each iteration. The propagation and reflection through a 2-D TLM

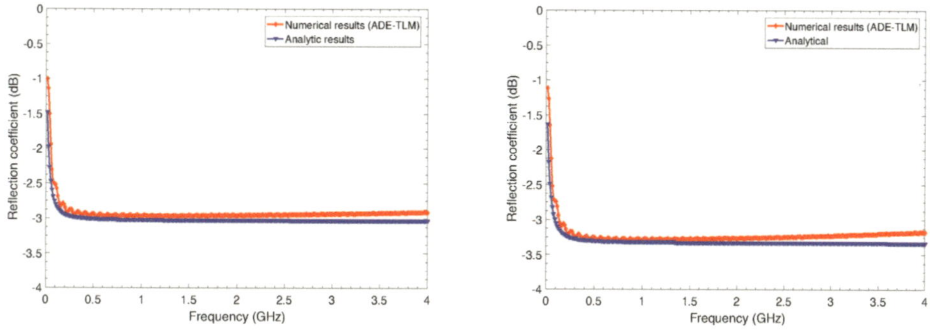

(a) (b)

Figure 11.
Reflection coefficient (dB) at normal incidence on an air/adenoma (a), air/Normal tyroidian tissue interface (b).

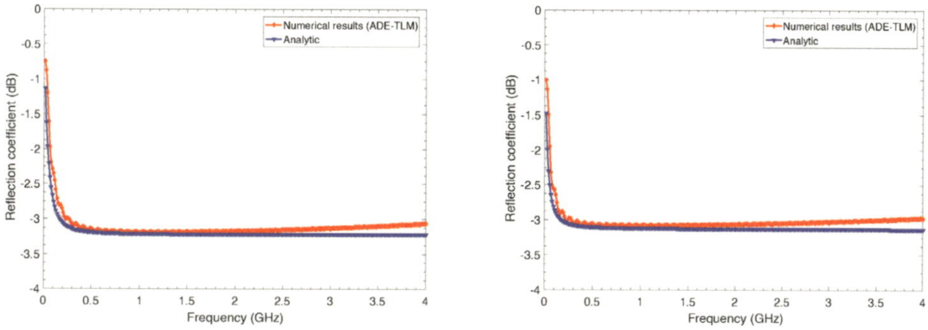

(a) (b)

Figure 12.
Reflection coefficient (dB) at normal incidence on an air.

grid modeling the incidence on an air/muscle interface is presented in **Figures 5** and **10** at different time steps, the interface is located at $z = 100$ in a 10 by 200 cells grid, **Figure 5(a)** shows the Gaussian wave before the arrival on the interface while **Figure 10(a)** and **(b)** depicts it after reflection and transmission. A more precise simulation that targeted the Dielectric properties of pathological thyroid tissue types including adenoma, thyroiditis and the properties of the healthy thyroid based on the results of the experimental study presented in a previous published work in [2], and for the healthy thyroid tissue [23], in both works experimental results are used to extract the 2 poles Cole-Cole model parameters. Based upon this parameters we conducted a simulation for the 4 tissue types, the results of this simulation are presented in **Figure 11(a)** and **(b)** for the air/adenoma and the air/normal thyroidian interface respectively, **Figure 12(a)** and **(b)** depicts the reflection coefficient on an air/ thyroyditis affected thyroidian tissue and an air/tumor respectively also in this case a perfect agreement over the working frequency is observed.

5. Conclusion

Numerical methods are an essential part of the modeling process, new propagation media with anomalous relaxation properties impose innovative modeling methods. An effective ADE-TLM formulation way to model electromagnetic wave propagation in biological tissues using Cole-Cole dispersion model. The fractional-order derivative in the Cole-Cole model is tackled by using a polynomial extrapolation of the polarization current. This approximation reduces considerably both the numerical cost of the simulation and its complexity since only two previous values of the *J* field are needed at each iteration. a Reimann derivation on the obtained polynomial extrapolation of the polarization current is then performed to solve the fractional order derivative. The presented results indicate the accuracy of our approach.

Conflict of interest

The authors declare no conflict of interest.

Author details

Mohammed Kanjaa*†, Otman El Mrabet† and Mohsine Khalladi†
University of Abdelmalek Essadi, Tetouan, Morocco

*Address all correspondence to: mohammedkanjaa@uae.ac.ma

† These authors contributed equally.

References

[1] Martellosio A, Pasian M, Bozzi M, Perregrini A, Mazzanti F, Svelto PE, et al. L: 0.5-50 ghz dielectric characterisation of breast cancer tissues. Electronics Letters. 2015;**51**(13):974-975

[2] Gavazzi S, Limone P, De Rosa G, Molinari F, and Vecchi G: Comparison of microwave dielectric properties of human normal, benign and malignant thyroid tissues obtained from surgeries: A preliminary study, Biomedical Physics Engineering Express, vol. 4, no. 4, p. 047 003, 2018.

[3] Ruvio G, Eaton-Evans J, Shahzad A, OHalloran M. Numerical evaluation of microwave thermal ablation to treat small adrenocortical masses. International Journal of RF and Microwave Computer-Aided Engineering. 2018;**28**(3):e21236

[4] Ley S, Schilling S, Fiser O, Vrba J, Sachs J, Helbig M. Ultra-wideband temperature dependent dielectric spectroscopy of porcine tissue and blood in the microwave frequency range. Sensors. 2019;**19**(7):1707

[5] Chakarothai J, Wake K, Watanabe S. Convergence of a single-frequency FDTD solution in numerical dosimetry. IEEE Transactions on Microwave Theory and Techniques. 2016;**64**(3): 707-714

[6] Debye P. Part i. dielectric constant. energy absorption in dielectrics with polar molecules. Transactions of the Faraday Society. 1934;**30**:679-684

[7] Cole KS, Cole RH. Dispersion and absorption in dielectrics i. alternating current characteristics. The Journal of Chemical Physics. 1941;**9**(4):341-351

[8] Rekanos IT, Yioultsis TV. Approximation of grunwald-letnikov fractional derivative for FDTD modeling of Cole-Cole media. IEEE Transactions on magnetics. 2014;**50**(2): 181-184

[9] Guo B, Li J, Zmuda H. A new FDTD formulation for wave propagation in biological media with Cole-Cole model. IEEE Microwave and Wireless Components Letters. 2006;**16**(12): 633-635

[10] Rekanos IT, Papadopoulos TG. An auxiliary differential equation method for FDTD modeling of wave propagation in Cole-Cole dispersive media. IEEE Transactions on Antennas and Propagation. 2010;**58**(11): 3666-3674

[11] Barba I. Cabeceira A C L, Panizo M, and Represa J: Modelling dispersive dielectrics in TLM method, International Journal of Numerical Modelling: Electronic Networks. Devices and Fields. 2001;**14**(1):15-30

[12] Samko SG, Kilbas AA, Marichev OI, et al. Fractional integrals and derivatives. Gordon and Breach Science Publishers, Yverdon Yverdon-les-Bains, Switzerland, 1993. In: vol. 1993

[13] Techniques in the fractional calculus," in The Fractional Calculus, ser. Mathematics in Science and Engineering, K. B. Oldham and J. Spanier, Eds., vol. 111, Elsevier, 1974, pp. 133–160.

[14] N. Engheta: n the role of fractional calculus in electromagnetic theory," IEEE Antennas and Propagation Magazine, vol. 39, no. 4, pp. 35–46, Aug. 1997.

[15] Engheta N. On fractional calculus and fractional multipoles in electromagnetism. IEEE Transactions on Antennas and Propagation. Apr. 1996;**44**(4):554-566

[16] Wharmby AW, Bagley RL. The application of the fractional calculus model for dispersion and absorption in

dielectrics in terahertz waves. International Journal of Engineering Science. 2015;**93**:1-12

[17] Jin H, Vahldieck R. Direct derivations of TLM symmetrical condensed node and hybrid symmetrical condensed node from maxwell's equations using centered differencing and averaging. IEEE Transactions on Microwave Theory and Techniques. Dec. 1994;**42**(12):2554-2561

[18] Christopoulos C, The Transmission-Line Modeling (TLM) Method in Electromagnetics. Morgan Claypool, 2006.

[19] Cabeceira ACL, Barba I, Grande A, Represa J. A 2D-TLM model for electromagnetic wave propagation in chiral media. Microwave and Optical Technology Letters. 2005;**46**(2):180-182

[20] Yaich MI, Khalladi M. The far-zone scattering calculation of frequency-dependent materials objects using the tlm method. IEEE Transactions on Antennas and Propagation. Nov. 2002; **50**(11):1605-1608

[21] Yaich MI, Kanjaa M, Adraoui S, Mounirh K, Khalladi M. An unsplit formulation of the 3D-PML absorbing boundary conditions for TLM-method in time domain. In: 2018 6th International Conference on Multimedia Computing and Systems (ICMCS). May 2018. pp. 1-5

[22] Taflove A, Hagness SC. Computational electrodynamics: the finite-difference time-domain method. Artech house. 2005

[23] Gabriel S, Lau RWand Gabriel C1996b The dielectric properties of biological tissues: III. Parametric models for the dielectric spectrum of tissues Phys. Med. Biol. 41 227193

Chapter 3

Averaged No-Regret Control for an Electromagnetic Wave Equation Depending upon a Parameter with Incomplete Initial Conditions

Abdelhak Hafdallah and Mouna Abdelli

Abstract

This chapter concerns the optimal control problem for an electromagnetic wave equation with a potential term depending on a real parameter and with missing initial conditions. By using both the average control notion introduced recently by E. Zuazua to control parameter depending systems and the no-regret method introduced for the optimal control of systems with missing data. The relaxation of averaged no-regret control by the averaged low-regret control sequence transforms the problem into a standard optimal control problem. We prove that the problem of average optimal control admits a unique averaged no-regret control that we characterize by means of optimality systems.

Keywords: optimal control, averaged no-regret control, electromagnetic wave equation, parameter depending equation, systems with missing data

1. Introduction

The research in the field of electromagnetism is set to become a vital factor in biomedical technologies. Those studies included several areas like the usage of electromagnetic waves for probing organs and advanced MRI techniques, microwave biosensors, non-invasive electromagnetic diagnostic tools, therapeutic applications of electromagnetic waves, radar technologies for biosensing, the adoption of electromagnetic waves in medical sensing, cancer detection using ultra-wideband signal, the interaction of electromagnetic waves with biological tissues and living systems, theoretical modeling of electromagnetic propagation through human body and tissues and imaging applications of electromagnetic.

Actually, the principal goal of the study is to control such electromagnetic waves to be compatible with some biomedical needs like X-rays in the framework of medical screening and wireless power transfer of electromagnetic waves through the human body [1] where we want to make waves closer to a desired distribution.

In this chapter, we consider a linear wave equation with a potential term $p(x, \sigma)$ supposed dependent on space variable x and real parameter $\sigma \in (0, 1)$, this term generally comprises the dielectric permittivity of the medium which has different properties and cannot be exactly presented, this is because of the difference or lack of knowledge of the physical properties of the material penetrated from the electromagnetic waves. The initial position and velocity are also supposed unknown.

In this study, we consider an optimal control problem for electromagnetic wave equation depending upon a parameter and with missing initial conditions. We use the method of no-regret control which was introduced firstly in statistics by Savage [2] and later by Lions [3, 4] where he used this concept in optimal control theory, and its related idea is "low-regret" control to apply it to control distributed systems of incomplete data which has the attention of many scholars [5–12], motivated by various applications in ecology, and economics as well [13]. Also, we use the notion of average control because our system depends upon a parameter, Zuazua was the first who introduced this new concept in [14].

The rest of this chapter is arranged as follows. Section 2, lists the definition of the problem we are studying. Section 3, is devoted to the study of the averaged no-regret control and the averaged low-regret control for the electromagnetic wave equation. Ultimately, we prove the existence of a unique average low-regret control, and the characterization of the average optimal is given in Section 4. Finally, we make a conclusion in Section 5.

2. Statement of the problem

Consider a bounded open domain Ω with a smooth boundary $\partial\Omega$. We set $\Sigma = (0, T) \times \partial\Omega$ and $Q = \Omega \times (0, T)$. We introduce the following linear electromagnetic wave equation depending on a parameter

$$\begin{cases} \dfrac{\partial^2 y}{\partial t^2} - \Delta y + p(x,\sigma)y = 0 & \text{in } Q \\ y = \begin{cases} v & \text{on } \Sigma_0 \\ 0 & \text{on } \Sigma\backslash\Sigma_0 \end{cases} & \\ y(x,0) = y_0(x); \ \dfrac{\partial y}{\partial t}(x,0) = y_1(x) & \text{in } \Omega \end{cases} \tag{1}$$

where $p \in L^\infty(\Omega)$ is the potential term supposed dependent on a real parameter $\sigma \in (0,1)$ presents the dielectric permittivity and permeability of the medium and such that $0 < \alpha_1 \leq p(x,\sigma) \leq \alpha_2$ *a.e.*in Ω, v is a boundary control in $L^2(\Sigma_0)$, $y_0 \in H_0^1(\Omega)$, $y_1 \in L^2(\Omega)$ are the initial position and velocity respectively, both supposed unknown. For all $\sigma \in (0,1)$, the wave Eq. (1) has a unique solution $y(v, y_0, y_1, \sigma)$ in $C([0,T]; H^1(\Omega)) \cap C^1([0,T]; L^2(\Omega))$ [15].

Denote by $g = (y_0, y_1) \in H_0^1(\Omega) \times L^2(\Omega)$ the initial data. We want to choose a control u independently of σ and g in a way such that the average state function y approaches a given observation $y_d \in L^2(Q)$. To achieve our goal, let' associate to (1) the following quadratic cost functional

$$J(v,g) = \left\| \int_0^1 y(v,g,\sigma)d\sigma - y_d \right\|^2_{L^2(Q)} + N\|v\|^2_{L^2(\Sigma_0)} \tag{2}$$

where $N \in \mathbb{R}_+^*$.

In this work, we aim to characterize the solution u of the optimal control problem with missing data given by

$$\inf_{v \in L^2(\Sigma_0)} J(v,g) \text{ subject to } (1) \tag{3}$$

independently of g and σ.

3. Averaged no-regret control and averaged low-regret control for the electromagnetic wave equation

A classical method to obtain the optimality system is then to solve the minmax problem

$$\inf_{v \in L^2(\Sigma_0)} \left(\sup_{g \in H_0^1(\Omega) \times L^2(\Omega)} (J(v,g)) \right), \tag{4}$$

but $J(v,g)$ is not upper bounded since $\sup_{g \in H_0^1(\Omega) \times L^2(\Omega)} (J(v,g)) = +\infty$. A natural idea of Lions [3] is to search for controls v such that

$$J(v,g) - J(0,g) \leq 0, \forall g \in H_0^1(\Omega) \times L^2(\Omega) \tag{5}$$

Those controls v are called averaged no-regret controls.

As in [16, 17], we introduce the averaged no-regret control defined by.

Definition 1 [1] We say that $v \in L^2(\Sigma_0)$ is an averaged no-regret control for (1) if v is a solution of

$$\inf_{v \in L^2(\Sigma_0)} \left(\sup_{g \in H_0^1(\Omega) \times L^2(\Omega)} (J(v,g) - J(0,g)) \right). \tag{6}$$

Let us start by giving the following important lemma.

Lemma 1 For all $v \in L^2(\Sigma_0)$ and $g \in G$ we have

$$J(v,g) - J(0,g) \quad = \quad J(v,0) - J(0,0) \tag{7}$$

$$-2\int_\Omega y_0(x) \int_0^1 \frac{\partial \zeta}{\partial t}(x,0) d\sigma dx + 2\int_\Omega y_1(x) \int_0^1 \zeta(x,0) d\sigma dx \tag{8}$$

where ζ is given by the following backward wave equation

$$\begin{cases} \dfrac{\partial^2 \zeta}{\partial t^2} - \Delta \zeta + p(x,\sigma)\zeta = \displaystyle\int_0^1 y(v,0,\sigma) d\sigma & \text{in } Q \\ \zeta = 0 & \text{on } \Sigma \\ \zeta(x,T) = 0, \ \dfrac{\partial \zeta}{\partial t}(x,T) = 0 & \text{in } \Omega \end{cases} \tag{9}$$

which has a unique solution in $C([0,T]; H_0^1(\Omega)) \cap C^1([0,T]; L^2(\Omega))$ [15].

Proof. It's easy to check that for all $(v,g) \in L^2(\Sigma_0) \times G$

$$J(v,g) - J(0,g) = J(v,0) - J(0,0) + 2\int_Q \left(\int_0^1 y(v,0) d\sigma \right) \left(\int_0^1 y(0,g) d\sigma \right) dx dt. \tag{10}$$

Use (9) and apply Green formula to get

$$\int_Q \left(\int_0^1 y(v,0)d\sigma\right)\left(\int_0^1 y(0,g)d\sigma\right)dxdt = \int_0^T\int_\Omega \left(\frac{\partial^2\zeta}{\partial t^2} - \Delta\zeta + p(x,\sigma)\zeta\right)\left(\int_0^1 y(0,g)d\sigma\right)dxdt \tag{11}$$

$$= -2\int_\Omega y_0(x)\int_0^1 \frac{\partial\zeta}{\partial t}(v,0)d\sigma dx + 2\int_\Omega y_1(x)\int_0^1 \zeta(v,0)d\sigma dx. \tag{12}$$

■

The no-regret control seems to be hard to characterize (see [11]), for this. reason we relax the no-regret control problem by making some quadratic perturbation as follows.

Definition 2 [17] We say that $u_\gamma \in L^2(\Sigma_0)$ is an averaged low-regret control for (1) if u_γ is a solution of

$$\inf_{v\in L^2(\Sigma_0)}\left(\sup_{g\in H_0^1(\Omega)\times L^2(\Omega)} \left(J(v,g) - J(0,g) - \gamma\|y_0\|^2_{H_0^1(\Omega)} - \gamma\|y_1\|^2_{L^2(\Omega)}\right)\right), \gamma > 0. \tag{13}$$

Using (9) the problem (13) can be written as

$$\inf_{v\in L^2(\Sigma_0)}\left(\begin{array}{c} J(v,0) - J(0,0) + \sup_{g\in G} 2\left(\int_\Omega y_1(x)\int_0^1 \zeta(v,\sigma)(x,0)d\sigma dx - \int_\Omega y_0(x)\int_0^1 \frac{\partial\zeta}{\partial t}(v,\sigma)(x,0)d\sigma dx\right. \\ \left. -\gamma\|y_0\|^2_{H_0^1(\Omega)} - \gamma\|y_1\|^2_{L^2(\Omega)}\right). \end{array}\right), \gamma > 0. \tag{14}$$

And thanks to Legendre transform (see [18, 19]), we have

$$\sup_{g\in G} 2\left(-\int_\Omega y_0(x)\int_0^1 \frac{\partial\zeta}{\partial t}(v,0)d\sigma dx + \int_\Omega y_1(x)\int_0^1 \zeta(v,0)d\sigma dx - \gamma\|y_0\|^2_{H_0^1(\Omega)} - \gamma\|y_1\|^2_{L^2(\Omega)}\right) \tag{15}$$

$$= \frac{1}{\gamma}\left\|\int_0^1 \frac{\partial\zeta(v,\sigma)}{\partial t}(x,0)d\sigma\right\|^2_{H_0^1(\Omega)} + \frac{1}{\gamma}\left\|\int_0^1 \zeta(v,\sigma)(x,0)d\sigma\right\|^2_{L^2(\Omega)}. \tag{16}$$

Then, the averaged low-regret control problem (9) is equivalent to the following classical optimal control problem

$$\inf_{v\in L^2(\Sigma_0)} J_\gamma(v) \tag{17}$$

where

$$J_\gamma(v) = J(v,0) - J(0,0) + \frac{1}{\gamma}\left\|\int_0^1 \frac{\partial\zeta(v,\sigma)}{\partial t}(x,0)d\sigma\right\|^2_{H_0^1(\Omega)} + \frac{1}{\gamma}\left\|\int_0^1 \zeta(v,\sigma)(x,0)d\sigma\right\|^2_{L^2(\Omega)}. \tag{18}$$

4. Characterizations

In the recent section, we aim to find a full characterization for the averaged no-regret control and averaged low-regret control via optimality systems.

Theorem 1.1 There exists a unique averaged low-regret control u_γ solution to (17), (18).

Proof. We have for every $v \in L^2(\Sigma_0) : J_\gamma(v) \geq -J(0,0)$, this means that (17), (18) has a solution.

Let $(v_n^\gamma) \in L^2(\Sigma_0)$ be a minimizing sequence such that

$$\lim_{n\to\infty} J_\gamma(v_n^\gamma) = J_\gamma(u_\gamma) = d_\gamma. \tag{19}$$

We know that

$$J_\gamma(v_n^\gamma) = J(v_n^\gamma, 0) - J(0,0) + \frac{1}{\gamma}\left\|\int_0^1 \frac{\partial \zeta(v_n^\gamma, \sigma)}{\partial t}(x,0)d\sigma\right\|^2_{H_0^1(\Omega)} + \frac{1}{\gamma}\left\|\int_0^1 \zeta(v_n^\gamma, \sigma)(x,0)d\sigma\right\|^2_{L^2(\Omega)} \leq d_\gamma + 1. \tag{20}$$

This implies the following bounds

$$\|v_n^\gamma\|_{L^2(\Sigma_0)} \leq C_\gamma, \tag{21}$$

$$\left\|\int_0^1 y(v_n^\gamma, 0, \sigma)d\sigma\right\|_{L^2(Q)} \leq C_\gamma, \tag{22}$$

$$\left\|\frac{\partial^2 \zeta_n}{\partial t^2} - \Delta \zeta_n + p(x,\sigma)\zeta_n\right\|_{L^2(Q)} \leq C_\gamma, \tag{23}$$

where C_γ is a positive constant independent of n. Moreover, by continuity w.r.t. data and (21) we get

$$\|y(v_n^\gamma, 0, \sigma)\|_{L^2(Q)} \leq C_\gamma. \tag{24}$$

By similar way an by using (22) we obtain

$$\|\zeta(v_n^\gamma, \sigma)\|_{L^2(0,T;H_0^1(\Omega))} \leq C_\gamma. \tag{25}$$

Then, from (21) we deduce that there exists a subsequence still denoted (v_n^γ) such that $v_n^\gamma \rightharpoonup u_\gamma$weakly in $L^2(\Sigma_0)$, and from (22) we get

$$y(v_n^\gamma, 0, \sigma) \rightharpoonup y_\gamma \text{ weakly in } L^2(Q). \tag{26}$$

Also, because of continuity w.r.t. data we have $y(v_n^\gamma, 0, \sigma) \rightharpoonup y(u_\gamma, 0, \sigma)$ weakly in $L^2(Q)$, by limit uniqueness $y_\gamma = y(u_\gamma, 0, \sigma)$ solution to

$$\begin{cases} \dfrac{\partial^2 y_\gamma}{\partial t^2} - \Delta y_\gamma + p(x,\sigma)y_\gamma = 0 & \text{in } Q, \\ y_\gamma = \begin{cases} u_\gamma \\ 0 \end{cases} & \begin{matrix}\text{on } \Sigma_0, \\ \text{on } \Sigma\backslash\Sigma_0,\end{matrix} \\ y_\gamma(x,0) = y_0(x); \dfrac{\partial y_\gamma}{\partial t}(x,0) = y_1(x) & \text{in } \Omega. \end{cases} \tag{27}$$

In other hand, use (24) and (22) to apply the convergence dominated theorem and, we have

$$\int_0^1 y(v_n^\gamma, 0, \sigma)d\sigma \rightharpoonup \int_0^1 y(u_\gamma, 0, \sigma)d\sigma \text{ weakly in } L^2(Q). \tag{28}$$

From (25) we deduce the existence of a subsequence still be denoted by $\zeta(v_n^\gamma, \sigma)(x, 0)$ such that

$$\zeta(v_n^\gamma, \sigma)(x, 0) \rightharpoonup \zeta(u_\gamma, \sigma)(x, 0) \text{ weakly in } H_0^1(\Omega), \tag{29}$$

then

$$\frac{\partial^2 \zeta_n}{\partial t^2} - \Delta\zeta_n + p(x, \sigma)\zeta_n \to \frac{\partial^2 \zeta_\gamma}{\partial t^2} - \Delta\zeta_\gamma + p(x, \sigma)\zeta_\gamma \, \text{in} \, D'(Q), \tag{30}$$

where $D(Q) = C_0^\infty(Q)$, and (23) leads to

$$\frac{\partial^2 \zeta_n}{\partial t^2} - \Delta\zeta_n + p(x, \sigma)\zeta_n \rightharpoonup f \;\text{ weakly in} \, L^2(Q). \tag{31}$$

Again, by limit uniqueness $\frac{\partial^2 \zeta_\gamma}{\partial t^2} - \Delta\zeta_\gamma + p(x, \sigma)\zeta_\gamma = \int_0^1 y(u_\gamma, 0, \sigma)d\sigma$ in $L^2(Q)$. Finally, ζ_γ is a solution to

$$\begin{cases} \frac{\partial^2 \zeta_\gamma}{\partial t^2} - \Delta\zeta_\gamma + p(x, \sigma)\zeta_\gamma = \int_0^1 y(u_\gamma, 0, \sigma)d\sigma & \text{in } Q, \\ \zeta_\gamma = 0 & \text{on } \Sigma, \\ \zeta_\gamma(x, T) = 0, \ \frac{\partial \zeta_\gamma}{\partial t}(x, T) = 0 & \text{in } \Omega. \end{cases} \tag{32}$$

The uniqueness of u_γ follows from strict convexity and weak lower semi-continuity of the functional $J_\gamma(v)$.■

After proving existence and uniqueness, we aim in the next theorem to give a full description to the average low-regret control for the electromagnetic wave equation.

Theorem 1.2 For all $\gamma > 0$, the average low-regret control u_γ is characterized by the following optimality system

$$\begin{cases} \frac{\partial^2 y_\gamma}{\partial t^2} - \Delta y_\gamma + p(x, \sigma)y_\gamma = 0, & \\ \frac{\partial^2 \zeta_\gamma}{\partial t^2} - \Delta\zeta_\gamma + p(x, \sigma)\zeta_\gamma = \int_0^1 y(u_\gamma, 0, \sigma)d\sigma, & \\ \frac{\partial^2 \rho_\gamma}{\partial t^2} - \Delta\rho_\gamma + p(x, \sigma)\rho_\gamma = 0, & \\ \frac{\partial^2 q_\gamma}{\partial t^2} - \Delta q_\gamma + p(x, \sigma)q_\gamma = \int_0^1 (\rho_\gamma + y(u_\gamma, 0, \sigma))d\sigma - y_d & \text{in } Q, \\ y_\gamma = \begin{cases} u_\gamma & \text{on } \Sigma_0 \\ 0 & \text{on} \Sigma \backslash \Sigma_0 \end{cases}, \zeta_\gamma = 0, & \\ \rho_\gamma = 0, q_\gamma = 0 & \text{on } \Sigma, \\ y_\gamma(0, x) = 0, \ \frac{\partial y_\gamma}{\partial t}(0, x) = 0, & \\ \zeta_\gamma(x, T) = 0, \ \frac{\partial \zeta_\gamma}{\partial t}(x, T) = 0, & \\ \rho_\gamma(x, 0) = -\frac{1}{\gamma}\int_0^1 \frac{\partial \zeta(u_\gamma)}{\partial t}(x, 0)d\sigma, \ \frac{\partial \rho_\gamma}{\partial t}(x, 0) = \frac{1}{\gamma}\int_0^1 \zeta(u_\gamma)(x, 0)d\sigma, & \\ q_\gamma(T, x) = 0, \ \frac{\partial q_\gamma}{\partial t}(T, x) = 0 & \text{in } \Omega, \end{cases} \tag{33}$$

with

$$u_\gamma = \frac{1}{N}\int_0^1 \frac{\partial p_\gamma}{\partial \eta} d\sigma \, \text{in} \, L^2(\Sigma_0). \tag{34}$$

Proof. From the first order necessary optimality conditions, we have

$$J^{\gamma'}(u_\gamma)(w) = \left(\int_0^1 y(u_\gamma, 0)d\sigma - y_d, \int_0^1 y(w,0)d\sigma\right)_{L^2(Q)} + N(u_\gamma, w)_{L^2(\Sigma_0)} +$$
$$\left(\frac{1}{\gamma}\int_0^1 \zeta'(u_\gamma)(0)d\sigma, \int_0^1 \zeta'(w)(0)d\sigma\right)_{L^2(\Omega)} + \left(\frac{1}{\gamma}\int_0^1 \zeta(u_\gamma)(0)d\sigma, \int_0^1 \zeta(w)(0)d\sigma\right)_{L^2(\Omega)} = 0 \tag{35}$$

for all $w \in L^2(\Sigma_0)$.

Now, let us introduce $\rho_\gamma = \rho(u_\gamma, 0)$ unique solution to

$$\begin{cases} \dfrac{\partial^2 \rho_\gamma}{\partial t^2} - \Delta \rho_\gamma + p(x,\sigma)\rho_\gamma = 0 & \text{in } Q, \\ \rho_\gamma = 0 & \text{on } \Sigma, \\ \rho_\gamma(x,0) = -\dfrac{1}{\gamma}\displaystyle\int_0^1 \frac{\partial \zeta(u_\gamma)}{\partial t}(x,0)d\sigma; \ \frac{\partial \rho_\gamma}{\partial t}(x,0) = \frac{1}{\gamma}\int_0^1 \zeta(u_\gamma)(x,0)d\sigma & \text{in } \Omega. \end{cases} \tag{36}$$

So that for every $w \in L^2(\Sigma_0)$, we obtain

$$J^{\gamma'}(u_\gamma)(w) = \left(\int_0^1 y(u_\gamma, 0) + \rho_\gamma d\sigma - y_d, \int_0^1 y(w,0)d\sigma - y(0,0)\right)_{L^2(Q)} + N(u_\gamma, w)_{L^2(\Sigma_0)}$$
$$= 0 \tag{37}$$

We finally define another adjoint state $q_\gamma = q(u_\gamma)$ as the unique solution of

$$\begin{cases} \dfrac{\partial^2 q_\gamma}{\partial t^2} - \Delta q_\gamma + p(x,\sigma)q_\gamma = \displaystyle\int_0^1 (\rho_\gamma + y(u_\gamma, 0, \sigma))d\sigma - y_d & \text{in } Q, \\ q_\gamma = 0 & \text{on } \Sigma, \\ q_\gamma(x,T) = 0, \ \dfrac{\partial q_\gamma}{\partial t}(x,T) = 0 & \text{in } \Omega. \end{cases} \tag{38}$$

Then (35) becomes

$$u_\gamma = \frac{1}{N}\int_0^1 \frac{\partial q_\gamma}{\partial \eta} d\sigma \, \text{in} \, L^2(\Sigma_0). \tag{39}$$

■

The previous Theorem gives a low-regret control characterization. For the no-regret control, we need to prove the convergence of the sequence of averaged low-regret control to the averaged no-regret control. Then, we announce the following Proposition.

For some constant C independent of γ, we have

$$\|u_\gamma\|_{L^2(\Sigma_0)} \leq C, \tag{40}$$

$$\left\| \int_0^1 y(u_\gamma, 0, \sigma) d\sigma \right\|_{L^2(Q)} \le C, \tag{41}$$

$$\| y(u_\gamma, 0, \sigma) d\sigma \|_{L^2(Q)} \le C, \tag{42}$$

$$\left\| \int_0^1 \frac{\partial \zeta(u_\gamma, \sigma)}{\partial t}(x, 0) d\sigma \right\|_{H^{-1}(\Omega)} \le C\sqrt{\gamma}, \tag{43}$$

$$\left\| \int_0^1 \zeta(u_\gamma, \sigma)(x, 0) d\sigma \right\|_{L^2(\Omega)} \le C\sqrt{\gamma}, \tag{44}$$

$$\|\rho_\gamma\|_{L^\infty(0,T;H_0^1(\Omega))} \le C, \tag{45}$$

$$\|q_\gamma\|_{L^\infty(0,T;H_0^1(\Omega))} \le C. \tag{46}$$

Proof. Since u_γ is a solution to (17) and (18), we get

$$J_\gamma(u_\gamma) \le J_\gamma(0), \tag{47}$$

then

$$J(u_\gamma, 0) + \frac{1}{\gamma}\left\| \int_0^1 \frac{\partial \zeta(u_\gamma, \sigma)}{\partial t}(x, 0) d\sigma \right\|^2_{L^2(\Omega)} + \frac{1}{\gamma}\left\| \int_0^1 \zeta(u_\gamma, \sigma)(x, 0) d\sigma \right\|^2_{L^2(\Omega)} \le J(0, 0), \tag{48}$$

this gives (40), (41), (42) and (43). The bound (43) follows by a way similar to (24).

From energy conservation property with (43) and (44).

$$E_{\rho_\gamma}(t) = \frac{1}{2\Omega}\left[\left| \frac{\partial \rho_\gamma}{\partial t} \right|^2 + |\nabla \rho_\gamma|^2 + q(x, \sigma)|\rho_\gamma|^2 \right] dx = E_{\rho_\gamma}(0) \le C, \tag{49}$$

we find (45).

To get q_γ estimates, just reverse the time variable by taking $s = T - t$ to find (46).

Lemma 2 The averaged low-regret control u_γ tends weakly to the averaged no-regret control u when $\gamma \to 0$.

Proof. From (40) we deduce the existence of a subsequence still be denoted u_γ such that

$$u_\gamma \rightharpoonup u \text{ weakly in } L^2(\Sigma_0), \tag{50}$$

let us prove u is an averaged no-regret control. We have for all $v \in L^2(\Sigma_0)$

$$J(u_\gamma, g) - J(0, g) - \gamma\|y_0\|^2_{H_0^1(\Omega)} - \gamma\|y_1\|^2_{L^2(\Omega)} \le \sup_{g \in H_0^1(\Omega) \times L^2(\Omega)} (J(v, g) - J(0, g)), \tag{51}$$

take $\gamma \to 0$ to find

$$J(u, g) - J(0, g) \le \sup_{g \in H_0^1(\Omega) \times L^2(\Omega)} (J(v, g) - J(0, g)), \tag{52}$$

i.e. is an averaged no-regret control. ■.

Finally, we can present the following theorem giving a full characterization the average no-regret control.

Theorem 1.3 The average no-regret control u is characterized by the following optimality system

$$\begin{cases} \frac{\partial^2 y}{\partial t^2} - \Delta y + p(x,\sigma) y = 0, \\ \frac{\partial^2 \zeta}{\partial t^2} - \Delta \zeta + p(x,\sigma)\zeta = \int_0^1 y(u,0,\sigma) d\sigma, \\ \frac{\partial^2 \rho}{\partial t^2} - \Delta \rho + p(x,\sigma)\rho = 0, \\ \frac{\partial^2 q}{\partial t^2} - \Delta q + p(x,\sigma) q = \int_0^1 (\rho + y(u,0,\sigma)) d\sigma - y_d & \text{in } Q, \\ y = \begin{cases} u & \text{on } \Sigma_0 \\ 0 & \text{on } \Sigma \backslash \Sigma_0 \end{cases}, \zeta = 0 \\ \rho = 0; q = 0 & \text{on } \Sigma, \\ y(0,x) = 0, \frac{\partial y}{\partial t}(0,x) = 0, \\ \zeta(x,T) = 0, \frac{\partial \zeta}{\partial t}(x,T) = 0, \\ \rho(x,0) = \lambda_1(x), \frac{\partial \rho}{\partial t}(x,0) = \lambda_2(x), \\ q(T,x) = 0, \frac{\partial q}{\partial t}(T,x) = 0 & \text{in } \Omega, \end{cases} \tag{53}$$

with

$$u = \frac{1}{N} \int_0^1 \frac{\partial p}{\partial \eta} d\sigma \text{ in } L^2(\Sigma_0), \tag{54}$$

and

$\lambda_1(x) = \lim_{\gamma \to 0} -\frac{1}{\gamma} \int_0^1 \frac{\partial \zeta(u_\gamma)}{\partial t}(x,0) d\sigma$ weakly in $H_0^1(\Omega)$,

$\lambda_2(x) = \lim_{\gamma \to 0} \frac{1}{\gamma} \int_0^1 \zeta(u_\gamma)(x,0) d\sigma$ weakly in $L^2(\Omega)$.

Proof. From (42) continuity w.r.t data, we can deduce that

$$y(u_\gamma, 0, \sigma) \rightharpoonup y(u,0,\sigma) \text{ weakly in } L^2(\Omega), \tag{55}$$

solution to

$$\begin{cases} \frac{\partial^2 y}{\partial t^2} - \Delta y + p(x,\sigma) y = 0 & \text{in } Q, \\ y = \begin{cases} u & \text{on } \Sigma_0, \\ 0 & \text{on } \Sigma \backslash \Sigma_0, \end{cases} \\ y(x,0) = y_0(x); \frac{\partial y}{\partial t}(x,0) = y_1(x) & \text{in } \Omega. \end{cases} \tag{56}$$

Again, by (41) and dominated convergence theorem

$$\int_0^1 y(u_\gamma, 0, \sigma)d\sigma \rightharpoonup \int_0^1 y(u, 0, \sigma)d\sigma \text{weakly in} L^2(\Sigma_0). \tag{57}$$

The rest of equations in (53) leads by a similar way, except the convergences of initial data $\rho(x, 0)$, $\frac{\partial \rho}{\partial t}(x, 0)$ which will be as follows.

From (43) and (44) we deduce the convergences of

$$-\frac{1}{\gamma}\frac{\partial \zeta(u_\gamma, \sigma)}{\partial t}(x, 0) \rightharpoonup \lambda_1(x) \text{ weakly in } H_0^1(\Omega), \tag{58}$$

and

$$\frac{1}{\gamma}\zeta(u_\gamma, \sigma)(x, 0) \rightharpoonup \lambda_2(x) \text{ weakly in } L^2(\Omega). \tag{59}$$

5. Conclusion

As we have seen, the averaged no-regret control method allows us to find a control that will optimize the situation of the electromagnetic waves with missing initial conditions and depending upon a parameter. The method presented in the paper is quite general and covers a wide class of systems, hence, we could generalize the situation to more control positions (regional, punctual, ...) and different kinds of missing data (source term, boundary conditions, ...).

The results presented above can also be generalized to the case of other systems which has many biomedical applications. This problem is still under consideration and the results will appear in upcoming works.

Acknowledgements

This work was supported by the Directorate-General for Scientific Research and Technological Development (DGRSDT).

Author details

Abdelhak Hafdallah* and Mouna Abdelli
Laboratory of Mathematics Informatics and Systems (LAMIS), University of Larbi Tebessi, 12002 Tebessa, Algeria

*Address all correspondence to: abdelhak.hafdallah@univ-tebessa.dz

References

[1] A. Hafdallah and A. Ayadi. Optimal control of electromagnetic wave displacement with an unknown velocity of propagation, International Journal of Control, DOI: 10.1080/00207179.2018.14581, 2018.

[2] Savage LJ. The Foundations of Statistics. 2nd ed. New York: Dover; 1972

[3] Lions JL. Contrôle à moindres regrets des systèmes distribués. C. R. Acad. Sci. Paris Ser. I Math. 1992;**315**:1253-1257

[4] Lions JL. No-regret and low-regret control. Economics and Their Mathematical Models, Masson, Paris: Environment; 1994

[5] Hafdallah A. On the optimal control of linear systems depending upon a parameter and with missing data. Nonlinear Studies. 2020;**27**(2):457-469

[6] Jacob B, Omrane A. Optimal control for age-structured population dynamics of incomplete data. J. Math. Anal. Appl. 2010;**370**:42 48

[7] Baleanu D, Joseph C, Mophou G. Low-regret control for a fractional wave equation with incomplete data. Advances in Difference Equations. 2016. DOI: 10.1186/s13662-016-0970-8

[8] Mophou G, Foko Tiomela RG, Seibou A. Optimal control of averaged state of a parabolic equation with missing boundary condition. International Journal of Control. 2018. DOI: 10.1080/00207179.2018.1556810

[9] Mophou G. Optimal for fractional diffusion equations with incomplete data. J. Optim.Theory Appl. 2015. DOI: 10.1007/s10957-015-0817-6

[10] Lions JL. Duality Arguments for Multi Agents Least-Regret Control. Paris: College de France; 1999

[11] Nakoulima O, Omrane A, Velin J. On the Pareto control and no-regret control for distributed systems with incomplete data. SIAM J. CONTROL OPTIM. 2003;**42**(4):1167 1184

[12] Nakoulima O, Omrane A, Velin J. No-regret control for nonlinear distributed systems with incomplete data. Journal de mathématiques pures et appliquées. 2002;**81**(11):1161-1189

[13] Choudhury PK, El-Nasr MA. Electromagnetics for biomedical and medici- nal applications. Journal of Electromagnetic Waves and Applications. 2015;**29**(17):2275-2277. DOI: 10.1080/09205071.2015.1103984

[14] Zuazua E. Averaged control. Automatica. 2014;**50**(12):3077 3087

[15] Kian Y. Stability of the determination of a coefficient for wave equations in an infinite waveguide. Inverse Probl. Imaging. 2014;**8**(3): 713-732

[16] Nakoulima O, Omrane A, Velin J. Perturbations à moindres regrets dans les systèmes dis- tribués à données manquantes. C. R. Acad. Sci. Ser. I Math. (Paris). 2000;**330**:801 806

[17] A. Hafdallah and A. Ayadi. Optimal Control of a thermoelastic body with missing initial conditions; International Journal of Control, DOI: 10.1080/00207179.2018.1519258,2018.

[18] Aubin JP. Lanalyse non linéaire et ses motivations économiques. Paris: Masson; 1984

[19] Attouch H, Wets RJB. Isometries for the Legendre-Fenchel transform. Transactions of the American Mathematical Society. 1986;**296**(1): 33-60

Section 2

Electromagnetic Wave Radiation and Scattering

Chapter 4

Using Electromagnetic Properties to Identify and Design Superconducting Materials

Fred Lacy

Abstract

Superconductors have a wide array of applications, such as medical imaging, supercomputing, and electric power transmission, but superconducting materials only operate at very cold temperatures. Thus, the quest to engineer room temperature superconductors is currently a hot topic of research. To accomplish this mission, it is important to have a complete understanding of the material properties that are being used to create these superconductors. Understanding the atomic and electromagnetic properties of the prospective materials will provide tremendous insight into the best choice for the materials. Therefore, a theoretical model that incorporates electromagnetic field theory and quantum mechanics principles is utilized to explain the electrical and magnetic characteristics of superconductors. This model can be used to describe the electrical resistance response and why it vanishes at the material's critical temperature. The model can also explain the behavior of magnetic fields and why some superconducting materials completely exclude magnetic fields while other superconductors partially exclude these fields. Thus, this theoretical analysis produces a model that describes the behavior of both type I and type II superconductors. Since there are subtle differences between superconductors and perfect conductors, this model also accounts for this distinction and explains why superconductors behave differently than perfect conductors. Therefore, this theory addresses the major properties associated with superconducting materials and thus will aid researchers in the pursuit of designing room temperature superconductors.

Keywords: conductivity, permittivity, permeability, resistivity, resonance, Schrödinger wave equation

1. Introduction

Superconductors are materials in which electricity can flow indefinitely because electrons can move through the material without losing energy. Superconductivity is a state in which a material's electrical properties or characteristics are altered when the temperature reaches a sufficiently low value. This temperature is known as the material's critical temperature and when the material falls below this temperature, two phenomena will result. One event that occurs is the electrical resistance drops to zero (or electric fields inside the material must vanish). The other outcome which takes place is that magnetic fields diminish inside the material. For some materials the magnetic field becomes zero (if a field exists in the material prior to the temperature change it becomes zero, and if a field is applied after the temperature change, it will not enter the material) [1–5].

Superconductors are known to exhibit zero electrical resistance or infinite electrical conductivity when the temperature of the superconductor reaches its critical temperature. Electrical conductivity is well described on the macroscale and can be modeled by Ohm's Law. Although macroscale theories usually are not appropriate on a microscale, Ohm's Law remains valid at the atomic level [6]. Bloch's theorem states that for perfect periodic lattices, conduction electrons will lose no energy and thus experience zero resistance [7]. However, since it is impractical to obtain a perfect lattice and since some materials exhibit superconductivity and others do not demonstrate this behavior (e.g., gold, silver, copper), additional theories and/or explanations are needed to explain the phenomenon of superconductivity.

Research is currently underway to develop superconducting materials at room temperature. Electrical engineers, physicists, and material scientists are engaged in this research to understand, identify, and create materials that exhibit superconducting properties. Thus, if superconductivity research can advance this science and develop room temperature materials, many industries will be revolutionized [8–13].

One goal of superconductivity research is to identify and/or create materials to transmit energy or information efficiently and dependably. Because electrical resistance in wires will lead to wasted energy and information loss during transmission, superconductivity research is important in unlocking these mysteries to explain the nature of these materials. Successfully accomplishing the goal in these research areas will revolutionize electrical power transmission and information technology [8–10].

Another major goal in superconductivity research is to develop room temperature superconductors for use in medical imaging. High intensity magnetic fields are needed for MRI and NMR imaging and in order to produce these large fields, superconducting wires are used. The image quality as well as the cost to operate these imaging systems depends upon the use of superconducting magnets. Creating and maintaining the necessary large magnetic fields requires a substantial amount of energy. If this energy is consumed by the resistance in the wires, then the quality and/or operating cost will suffer. Achieving the goal of creating room temperature superconductors will further reduce the cost to operate and transform the medical industry [11–13].

Regarding electrical field properties, electrical resistivity is a fundamental material property that leads to electrical resistance in a material. Electrical resistance is a function of the dimensions of the material that is being utilized or analyzed, whereas electrical resistivity is based on atomic interactions with conduction electrons (thus resistivity is independent of the material dimensions). Electrical resistivity is a function of temperature and a material's resistivity will generally decrease as its temperature decreases [14]. When a material's temperature decreases, conduction electrons will interact less with lattice atoms and therefore they will lose less energy. When a superconductor enters the superconducting state and its electrical resistance vanishes, electrons will flow through that material unimpeded and current flows indefinitely [15].

Regarding magnetic field properties, the Meissner effect is a condition that results in magnetic fields vanishing inside the interior of superconductors (provided that the magnetic field is small). For these small magnetic fields, whether the field is present before or after the material is cooled below its critical temperature and whether the material is a type I or type II superconductor, that magnetic field will be expelled from the superconductor's interior. If the magnetic field increases and has a value between the upper and lower critical fields, type II superconductors will enter a mixed state in which a portion of the magnetic field will penetrate the superconductor's interior [1–5].

To truly understand why some materials are superconductors and why others are not, and to advance research in superconductivity, theoretical models are needed. Various microscopic or atomic theories have been developed to explain superconductivity [16]. Thermodynamic theories have been developed to explain certain aspects of superconductivity behavior, however, there have been inconsistencies in these theories or explanations [17]. Researchers understand that even though theories may be incorrect or partially correct, those theories can push the science forward and provide some insight into a material's behavior. For example, the BCS theory and the London equations are two of the most successful theories on superconductivity, but they have limitations in certain aspects of their explanations. In particular, the BCS theory was developed several decades before high temperature superconductors were discovered. As a result, it does well in describing the behavior of type I materials, but it is inadequate in explaining how high temperature superconductors operate [18–20]. Furthermore, the BCS theory is imperfect and insufficient in explaining several fundamental properties of superconducting materials such as the Meisner effect [18]. Again, a theory does not have to explain every aspect of a material's behavior, but if it cannot explain the fundamental properties, then it is inadequate. Therefore, a comprehensive theory that explains the electromagnetic properties of all superconducting materials is needed.

There is not a complete theory on superconductivity that explains the electromagnetic properties of type I and type II superconducting materials. Because these previously developed models are incomplete, one model may explain one aspect, but it cannot address another. A given theory can handle certain characteristics of superconductors, but it fails in other areas. So, if scientific research is going to make significant advances, a comprehensive theory is needed.

A general theoretical model has been developed to explain the relationship between conductivity and temperature [21, 22]. This model used atomic analysis and solid state physics principles to develop the theory and explain why electrical conductivity is dependent on temperature. The relationship between conductivity and temperature is first derived and then its accuracy is demonstrated through comparisons to known linear responses from platinum and nickel. Again, a model does not have to be entirely correct in order to be useful and move the science forward. This aforementioned model does not and did not intend to account for superconducting effects. Therefore, the model presented herein will provide missing pieces of the puzzle and demonstrate that it is sufficient to characterize properties of superconductors.

The theory presented here accounts for the electromagnetic properties of both type I and type II superconductors. The theoretical model has been obtained by using quantum mechanics and analyzing conduction electron interactions with atoms in a lattice. This analysis is then used to specify how the electrical resistance of a material will respond. Analysis reveals that certain conditions will allow electrons to move through a material without interference. Under these conditions, the material will act as a superconductor. This theoretical model will then be used to explain the Meisner effect or the response of superconducting materials to external magnetic fields. Since type I and type II superconducting materials display different characteristics, the model will be used to explain this difference. The theory can be used to explain the difference between perfect conductors and superconductors. Finally, the frequency spectrum of a generic material's dielectric response is analyzed to demonstrate the feasibility of this model in explaining superconductivity.

2. Why a new resistance model is needed

To determine the electrical resistivity of a material (or the electrical resistance since resistivity and resistance are proportional), Matthiessen's rule states that the total resistivity is comprised of the sum of the individual resistivities associated with electron interactions with lattice phonons as well as lattice imperfections [1]. In equation form the resistivity is $\rho = \rho_p + \rho_i$ and thus the equivalent electrical resistance equation is

$$R = R_p + R_i \tag{1}$$

where R_p is the resistance that arises due to phonons and is a function of temperature, and R_i is the resistance that arises due to lattice imperfections (e.g., defects, impurities, grain boundaries, etc.) and is independent of temperature. In essence, Matthiessen's rule is a 'series' approach to electrical resistance, but this approach alone cannot account for the electrical resistance of superconductors since all resistance terms would have to become zero in order for the total electrical resistance to become zero. **Figure 1** shows the individual resistances and the total or sum of these resistance for a typical material. This figure demonstrates that a series approach alone cannot explain electrical resistance of superconductors. As a result, modifications must be made to Matthiessen's rule or a different approach has to be taken to account for superconducting effects in materials. Rather than using a series concept to analyze and explain electrical resistance, a general approach using parallel concepts will be used.

The two-fluid model and the resistively shunted junction have been used to describe superconductivity and it incorporates a parallel approach to phenomenologically describe electrical resistance [23, 24]. These approaches use a 'regular' channel to represent normal or non-superconducting electrons and a 'superconductor' channel to represent superconducting or Cooper-paired electrons. It is presupposed that the Matthiessen series model can be incorporated into the regular channel and therefore the regular channel can model the non-superconducting response of materials. However, since Cooper-paired electrons cannot explain the behavior of high-temperature superconducting materials, these models need to be modified to include or reflect the underlying physics as well as the behavior of all types of superconductors.

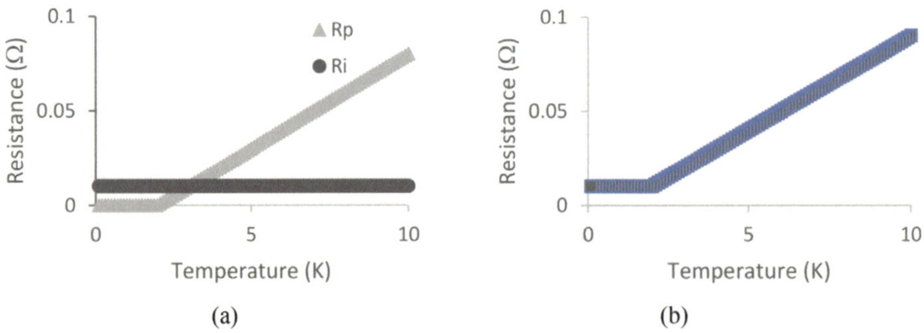

Figure 1.
(a) Example of electrical resistances for a material associated with conduction electron interaction with lattice atoms (phonons) R_p, as well as conduction electron interaction with lattice imperfections R_i. (b) Example of the total resistance using Matthiessen's rule in which the resistance is the sum of the individual resistances. Using Matthiessen's rule makes it impossible to explain why electrical resistance vanishes in superconductors.

3. Electrical resistance analysis

Information about this model has been developed and published [21, 22, 25, 26]. Since the conduction electrons shown in **Figure 2** travel down two different paths, they will encounter two different electrical resistances. Resistance R_1 will represent the resistance that conduction electrons experience if they are not in the space or pathway of atoms. Resistance R_2 will represent the resistance that conduction electrons experience if they are in the path of atoms. Since these resistors are in parallel, they have an equivalent resistance represented by R as given in Eq. (2).

$$\frac{1}{R}=\frac{1}{R_1}+\frac{1}{R_2} \tag{2}$$

To develop and understand the theoretical model, consider a generic atomic lattice shown in **Figure 2**. The model is a two-dimensional lattice structure with periodic atoms and conduction electrons that travel through the lattice. An electron will either travel through a path that contains atoms, or it will not. Atoms will vibrate at a rate that is a function of temperature. These vibrations will enlarge or expand the space that contains atoms or conversely, as temperature decreases, the space between atoms will increase. A higher proportion of conduction electrons in the space between atoms will lead to higher electrical conductivity. By analyzing a unit cell within the lattice, an equation can be obtained that represents the response for the entire lattice.

This model will yield an equation that describes the relationship between conductivity (or resistivity) and temperature. This model utilizes the concept that when electrons are in the space or path containing atoms, these atoms will always impede the flow of the electrons. In the case of superconducting materials at temperatures below their critical temperature, this will not be the case. So, in order to address superconductivity with this model, additional elements will be factored into the analysis.

When conduction electrons are in the region where they will only directly encounter other electrons, the electrical resistivity in this region will be temperature dependent and will have a response that linearly decreases as temperature decreases. A typical example of this response is shown in **Figure 3** where the

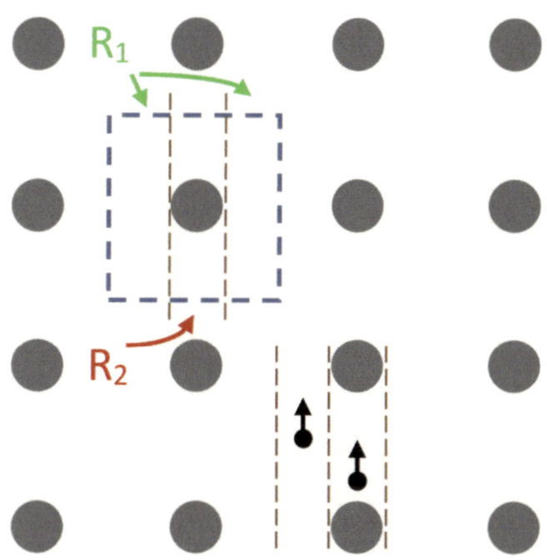

Figure 2.
Illustration of the atomic lattice showing electrons (small circles) and atoms (large circles). In a non-superconducting state, an electron will experience very little resistance (leftmost electron) if it travels between atoms (resistance R_1) and much resistance (rightmost electron) if it travels in the path of atoms (resistance R_2). These are the two cases that can occur within a unit cell (dotted box).

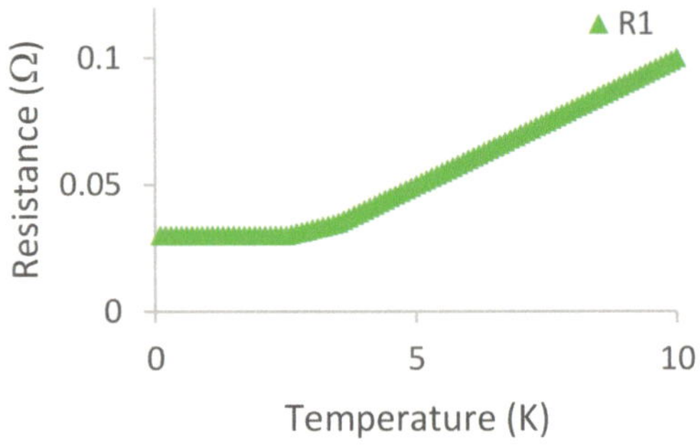

Figure 3.
Example of electrical resistance as a function of temperature for conduction electrons that travel in pathways between atoms and will not directly contact atoms. This graph represents the resistance of resistor R_1.

resistivity ρ and resistance R_1 generally decreases with decreasing temperature. Because of lattice defects, grain boundaries, and impurities, resistance R_1 will not reach zero when the temperature is zero, but it will have a residual value. This represents the resistance in the region where atoms do not exist regardless of whether the material is superconducting or non-superconducting.

Resistance R_2 represents electrons that will travel in regions where they will encounter atoms. This resistance will have two different responses depending upon whether the material is a superconductor or not. If the material is a superconductor, then at some critical temperature, the atoms will offer no resistance or they will be invisible to conduction electrons, but above this critical temperature, there will be a non-zero temperature dependent response. An example of this response is shown in **Figure 4a**. For non-superconducting materials, the electrical resistance R_2 will not become zero like it does for superconductors at the critical temperature. The resistance will have some temperature dependent response and have some non-zero value when the temperature reaches zero. An example of this response is shown in **Figure 4b**.

This evaluation provides an analysis of the resistances that can (and will) be used in the theoretical model. To fully incorporate superconducting effects into the theoretical model, the interaction between the atoms and conduction electrons must be examined. This will provide details that explain when and why certain materials become superconductors. The analysis of the conduction electron and atom interaction will be accomplished by utilizing the Schrödinger wave equation and quantum mechanics.

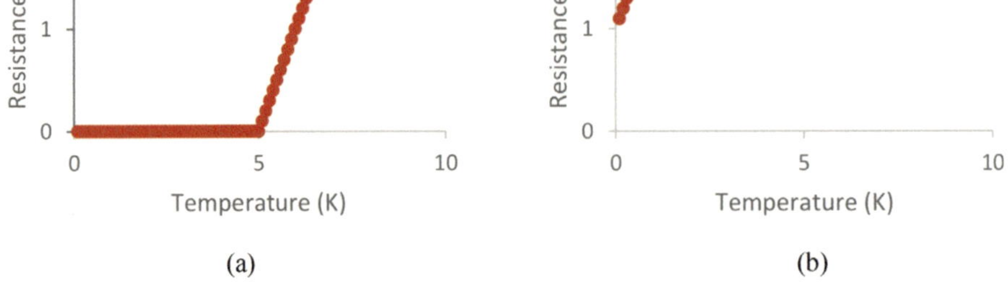

Figure 4.
Example of electrical resistance as a function of temperature for conduction electrons traveling in the pathway of atoms (a) in a superconducting material [type I] and (b) in a non-superconducting material. These graphs represent the resistance of resistor R_2.

4. Atomic theory analysis

Through the wave particle duality theorem, it is understood that objects behave like waves and like particles. Thus, atomic particles will display both particle-like properties as well as wave-like properties. To highlight this theorem, electrons are indeed particles with electrical charge and mass, but they are also waves that travel with a wavelength and frequency [1]. Traveling electrons can be characterized by the Schrödinger wave equation as shown in Eq. (3).

$$-\frac{\hbar^2}{2m}\frac{\partial^2\psi(x,t)}{\partial x^2}+V\psi(x,t)=i\hbar\frac{\partial\psi(x,t)}{\partial t} \tag{3}$$

In this equation, V represents the potential energy that a traveling electron will encounter at a particular time and location. Because many electrical charges exist in this model, the overall potential energy can be represented by the sum of the potentials associated with the charges that a traveling electron will encounter. So

$$V=\sum\frac{qQ}{4\pi\varepsilon x} \tag{4}$$

where q is the charge of a traveling electron, Q represents the charge of a nearby object (e.g., nucleus or another electron), ε is the permittivity or dielectric constant, and x is the distance between the traveling electron and nearby object.

Because there are many charges in the system, the exact solution to the Schrödinger wave equation will be extremely complex. However, the general solution to the wave equation, or Eq. (3) is given by

$$\psi(x,t)=Ae^{j(kx-\omega t)}u_k(x) \tag{5}$$

where k is the wavenumber, ω is the frequency of the traveling electron wave, and $u_k(x)$ is the Bloch function and has the periodicity of the lattice [1]. If the potential energy, V, is zero, then the electron will be a free particle and thus the wavenumber, k, will be a real number. This means that the wave can be modeled with a sinusoidal response and thus the electron will travel unimpeded.

Thus, if and when the permittivity becomes large enough, the potential energy will become sufficiently small and thus as conduction electrons travel through the lattice, they will not be attenuated. Therefore, if or when the permittivity becomes infinitely large, atoms will offer no resistance to moving electrons.

5. Magnetic field analysis

The electrical resistance response of a material is a major characteristic to determine if it is a superconductor. However, the response of a material to magnetic fields is also a major factor in determining if the material is a superconductor as well as what type of superconductor the material is. Therefore, magnetic field analysis is needed and must be incorporated into the theoretical model. To integrate a material's magnetic field response into the model, electromagnetic field theory will be considered.

Maxwell's equations can be used to understand how electromagnetic fields interact with matter [27]. Information such as how much of an electromagnetic wave will be transmitted or reflected at an interface between two different materials, and a wave's velocity, frequency, and wavelength in a material can be determined by analyzing Maxwell's equations. To gain a clear understanding of why the Meissner effect occurs in superconductors, Ampere's law (one of Maxwell's equations), will be used. In differential equation form

$$\nabla x\overline{B} = \mu\left(\overline{J} + \varepsilon\frac{d\overline{E}}{dt}\right) \tag{6}$$

where B is the magnetic flux, μ is the permeability of the material, ε is the permittivity of the material, J is the current density (and $\overline{J} = \sigma\overline{E}$), and E is the electric field. Electric fields and magnetic fields are related, and Eq. (6) shows that a magnetic field can be produced from an electric current and from a time changing electric field.

In addition to using Ampere's law as given in Eq. (6), the theoretical model will incorporate other electromagnetic equations and concepts because of the interdependence of the electric and magnetic fields. Specifically, Maxwell-Faraday's equation will be used or equivalently in differential equation form, $\nabla x\overline{E} = -\frac{d\overline{B}}{dt}$, and the force on charged particles due to external fields will be used or equivalently $\overline{F} = q\left(\overline{E} + \overline{v}\,x\,\overline{B}\right)$.

Eq. (6) reveals that the magnetic field is related to the electric field (or the time derivative of the electric field), but this equation also illustrates that this relationship is dependent upon the material's permittivity. Since a material's permittivity will behave as outlined and described previously, the main properties that characterize superconductors, zero electrical resistance and zero interior magnetic field are interrelated and therefore will be coupled.

6. Electrical resistance results

To demonstrate that the theoretical model functions properly and characterizes the behavior of superconducting materials, the resistance models, the parallel resistor concept, and the Schrödinger wave equation (and its solutions), will be used. Consequently, the model will be validated by demonstrating that it accurately describes the electrical resistance of both superconducting and non-superconducting materials. Furthermore, the distinction between the resistance of type I and type II superconductors will be made.

First the case for superconducting materials is considered. Resistances R_1 and R_2 are displayed on the same graph as a function of temperature as shown in **Figure 5a**. These curves are representative of the two resistances that would occur for superconductors. The theoretical model is represented by parallel resistors, so when these values are combined using the parallel resistor equation, the result is shown in **Figure 5b**. It is seen that the equivalent resistance has a linearly decreasing slope until the temperature reaches the transition temperature T_C. When it reaches this temperature, the resistance then abruptly goes zero. This is typical for the electrical resistance response of a superconducting material [1–5].

Next, the case for non-superconducting materials is considered. Resistances R_1 and R_2 are displayed on the same graph as a function of temperature as shown in **Figure 6a**. These curves are symbolic of the two resistance types that would occur for non-superconductors. Again, the theoretical model is represented by parallel resistors, so when these values are combined using the parallel resistor equation, the result is shown in **Figure 6b**. It is seen that the equivalent resistance resembles resistance R_1. This result is accurate and expected because resistance R_2 is always much larger than resistance R_1 and since parallel resistors will resemble the smaller

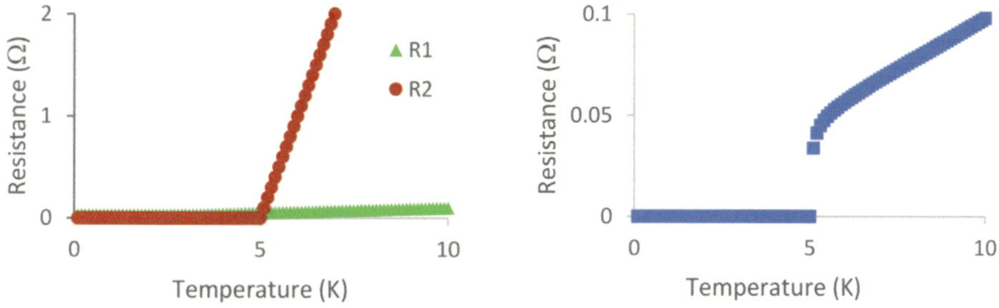

(a) (b)

Figure 5.
Example of electrical resistances for a type I superconducting material. (a) Graph showing the resistance of electrons traveling in the 'gaps' (represented by resistor R_1) and the resistance of electrons traveling in the pathway of atoms (represented by resistor R_2). (b) Graph showing the total resistance of a superconducting material when the two resistors R_1 and R_2 are combined in parallel. There is no resistance from the atoms when the material reaches its transition temperature.

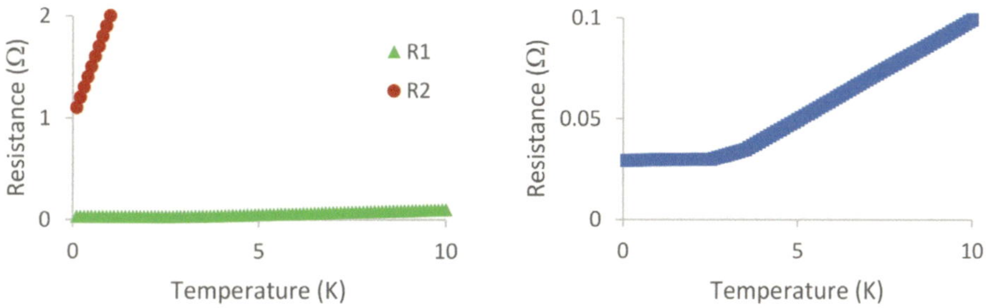

(a) (b)

Figure 6.
Example of electrical resistances for a non-superconducting material. (a) Graph showing the resistance of electrons traveling in the 'gaps' (represented by resistor R_1) and the resistance of electrons in the pathway of atoms (represented by resistor R_2). (b) Graph showing the total resistance of a non-superconducting material when the two resistors R_1 and R_2 are combined in parallel. The resistance of a non-superconductor will not reach zero regardless of the material's temperature.

resistor value, the total resistance will approximately equal R_1. As a result, when the resistance of non-superconductors is analyzed over a wide temperature range, there is no abrupt change in resistance because R_2 does not have a transition temperature. Therefore, conduction electrons will never experience zero resistance when they encounter atoms in this material. This is typical for the electrical resistance response of a non-superconducting material [1–5].

The previous analysis focused on revealing the electrical properties of type I superconductor materials. But the same theoretical model can be used to explain the behavior of type II superconductors materials. It is noted that there is a subtle difference between the electrical resistance response of type I materials and type II materials. At the transition temperature, the resistance of type I materials has a sharper transition to zero (i.e., the transition occurs over a small temperature range) whereas type II materials exhibit a slower transition (i.e., the transition occurs over a large temperature range) [28]. This difference in the transition temperature range can be explained using the following analysis.

Resistance R_1 (which is the resistance seen by conduction electrons that will not directly encounter atoms) will be the same for type I and type II materials. However, resistance R_2 (which is the resistance seen by the conduction electrons that directly interact with lattice atoms) for type II materials will be different than the

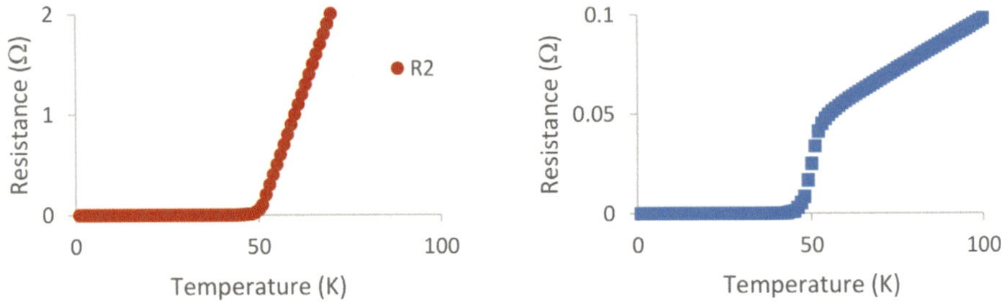

(a) (b)

Figure 7.
*Example of electrical resistances for a type II superconducting material. (a) Graph of electrical resistance R_2 for electrons traveling in the pathway of atoms and (b) graph of the total electrical resistance for a type II superconducting material. The total electrical resistance is obtained by combining resistances R_2 (as shown in this figure) and R_1 (as shown in **Figure 2**) in a parallel manner.*

resistance of type I materials. The difference is that R_2 (or in essence the permittivity of the atoms), will have a response that is not linear near the transition temperature. Whereas type I materials will have a resistance with a sharp response as the temperature approaches the critical temperature, type II materials will have a resistance that gradually approaches zero near the critical temperature. The type II material response is shown in **Figure 7a**. When the resistance shown in **Figure 7a** is used to represent resistance R_2, the equivalent resistance will have the response shown in **Figure 7b**. This equivalent resistance response is obtained experimentally for typical type II materials [28].

7. Magnetic field results

The theoretical model will now be validated by demonstrating that it accurately describes the response of both superconducting and non-superconducting materials to applied magnetic fields. When superconducting materials are above their critical temperature they will behave as other materials and thus external magnetic fields can exist on the interior of the superconductor. Upon cooling the superconductor to a temperature below its critical temperature, regardless of whether a magnetic field is applied before or after cooling, type I superconductors will have no interior magnetic field [1–5]. However, the manner in which the magnetic field is excluded from the interior of the superconductor in these two cases is different. These two cases of applying the magnetic field before and after cooling will be analyzed. To simplify this analysis, it is assumed that there is no initial current flowing through the superconductor, and thus current density J is zero (it is noted that this assumption will not affect the magnetic field analysis or results).

First, the case of applying the magnetic field after cooling a type I superconductor below its critical temperature is examined. In the area where atoms are located, the permittivity ε will be infinite and initially since there are no fields, the electric field is zero and dE/dt is zero. Now, when a magnetic field is applied, this field changes in space and time. The change goes from zero to a value that is non-zero and finite. So, ∇xB will be non-zero and thus dE/dt will become non-zero and finite and as a result, $(\varepsilon)(dE/dt)$ will be infinite. Since Ampere's law as shown in Eq. (6) must be true, $(\mu)(\varepsilon)(dE/dt)$ must be finite and this will only be true if the permeability $\mu = 0$. Since μ is zero, there will be no magnetic flux in the material. In essence, conduction electrons on the material's surface will circulate according to

$\overline{F} = q\overline{v}\ x\ \overline{B}$ and produce magnetic fields that will cancel the external field. Thus, there will be no net magnetic flux inside the material (it is noted that the magnetic field will decay exponentially according to the London penetration depth [1, 2, 29]). When the external magnetic field is removed, the conduction electrons stop circulating and return to their normal motion. The top half of **Figure 8** illustrates the magnetic field's behavior when a type I superconductor is cooled before the magnetic field is applied. Note that if the external magnetic field becomes too large, it will stretch and alter the atoms in the material, the surface current will be affected, and the permittivity ε of the atoms will no longer be infinite. As a result, the permeability μ will no longer be zero and therefore, the magnetic field will be able to penetrate the material.

Next the case of applying the magnetic field before cooling a type I superconductor below its critical temperature is examined. Initially the permittivity ε, the permeability μ, and the magnetic flux B inside the material will all be non-zero and finite. Then when the material is cooled and reaches the transition or critical temperature, the permittivity will change from finite to infinite. As a result, conduction electrons will circulate to counteract the applied field according to $\overline{F} = q\overline{v}\ x\ \overline{B}$. These circulating charges will reduce the external magnetic field. As a result of the changing magnetic flux, a changing electric field E will be produced, and therefore dE/dt will become non-zero. Since $\nabla x B$ will now be non-zero and finite, then the term $(\mu)(\varepsilon)(dE/dt)$ must be finite. This will be true if the permeability becomes zero or $\mu = 0$. If μ is zero, there will be no magnetic flux in the material, so the field will get expelled from the interior of the superconductor. Again, when the external magnetic field is removed, the conduction electrons stop circulating and return to their normal motion. The bottom half of **Figure 8** illustrates the magnetic field's behavior when a type I superconductor is cooled after the magnetic field is applied. Again, note that if the external magnetic field becomes too large, it will stretch and alter the atoms and subsequently surface currents and superconducting properties will be destroyed (thus the electrical resistance will no longer be zero and a magnetic field can penetrate the material).

The previous analysis focused on revealing the magnetic properties or magnetic field response of type I superconductor materials. The model (along with the same analysis) is not only capable of explaining why type II superconductors exhibit their response, but it is able to explain the behavior of type II materials. Type II superconductors have two critical magnetic fields: a lower field T_{CL} and a higher field T_{CH}. If the applied magnetic field is less than T_{CL}, the type II material behaves as a type I material and excludes all of the magnetic flux from its interior. If the applied

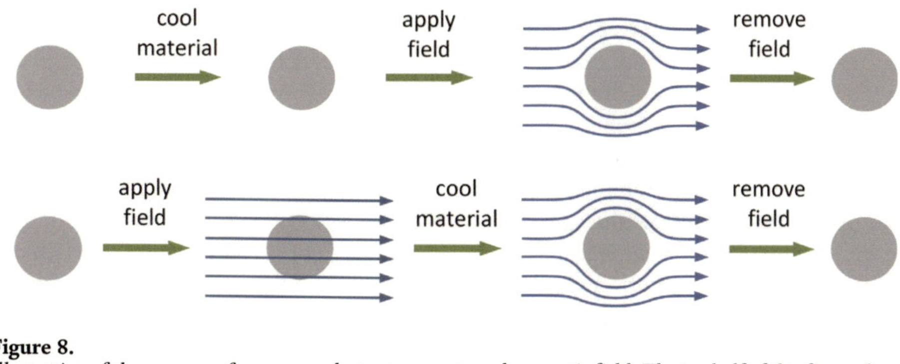

Figure 8.
Illustration of the response of a superconductor to an external magnetic field. The top half of this figure shows the response of the superconductor if the material is cooled before the field is applied and the bottom half of the figure shows the response if the field is applied before the material is cooled. Once the material is cooled, there will be no magnetic field inside the superconductor.

magnetic field is greater than T_{CH}, the type II material also behaves like type I materials and will allow all of the magnetic field to penetrate its interior. However, when the applied field is between T_{CL} and T_{CH} only a portion of the external field will penetrate type II materials. This is the mixed state in which superconducting and non-superconducting regions exist in the material [1–5]. Based on the state of the surface current and the state of the permittivity of the atoms in the material, the three different states of type II superconductors (i.e., superconducting, mixed, and normal) can be explained.

In the presence of a magnetic field, the atoms in the material will be stretched and altered. As long as the external magnetic field is smaller than T_{CL}, the atoms are not altered significantly. So, surface currents will exist, and the permittivity of the atoms will be infinite, and the material will exist in the superconducting state. This response is similar to type I materials, so no additional analysis is necessary.

The permittivity of the atoms in the material will vary as a function of the magnetic field (as well as temperature). So, when the external magnetic field exceeds T_{CL}, the permittivity of the atoms will be affected. Thus, the permittivity ε will not be infinite and based on Ampere's law, the permeability μ will no longer be zero. So, a magnetic field will be able to exist in the material. The surface currents will remain until the magnetic field reaches the upper critical field T_{CH}. The surface currents will block most of the magnetic field but some of the external field will penetrate through the surface. Since the permeability is not zero, the magnetic field will be able to exist in the interior of the material. This creates a pattern of superconducting and non-superconducting regions and forms an Abrikosov lattice [30]. This is the mixed state.

Finally, if the magnetic field is continually increased and reaches T_{CH}, the atoms in the material become stretched to the point where they interfere with all of the surface currents. Since the surface currents will not be able to flow unimpeded and since the permittivity of the atoms will not be infinite, the material moves from the mixed state to a non-superconducting state.

8. Superconductors vs. perfect conductors

If a model that explains superconductivity is going to be completely correct, it must also be able to explain why superconductors are different from perfect conductors. This theoretical model can explain this difference. The main difference between a perfect conductor and a superconductor is in the response to a magnetic field that is applied before the material is cooled to the transition temperature. A superconductor will exclude the magnetic field when it reaches the transition temperature whereas a perfect conductor will allow the magnetic field to remain [28, 31]. The magnetic field analysis for superconductors has been performed using the theoretical model, and the model explains the behavior. A similar analysis will be utilized to explain the behavior of perfect conductors.

A perfect conductor is defined as a material that has no resistivity or equivalently zero conductivity when the material is below the transition temperature. So, to compare the perfect conductor to a superconductor, $\sigma = \infty$ for the perfect conductor, but $\varepsilon = \infty$ for superconductors. Although the perfect conductor will have infinite conductivity, its other electromagnetic parameters, ε and μ, remain normal. Since $\sigma = \infty$ and $J = \sigma E$, the electric field E must vanish in the material to keep the current density J finite. As a result, dE/dt must be zero. Furthermore, the term $(\mu)(\varepsilon)(dE/dt)$ must be zero since the first two terms are finite and dE/dt will be zero. Ampere's law requires that ∇xB and $(\mu)(\varepsilon)(dE/dt)$ must be equal, so ∇xB must be zero and therefore B cannot change spatially. Thus, the perfect conductor will

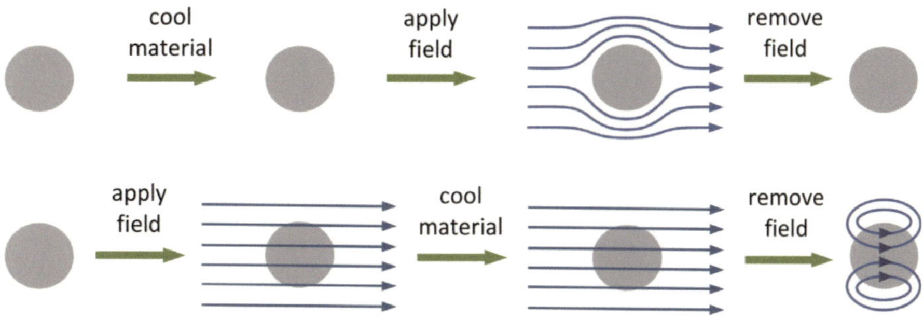

Figure 9.
Illustration of the response of a perfect conductor to an external magnetic field. The top half of this figure shows the response of the perfect conductor if the material is cooled before the field is applied and the bottom half of the figure shows the response if the field is applied before the material is cooled. Regardless of when cooling occurs, the material maintains the magnetic field that it had when the external field is removed.

contain the same amount of magnetic flux before and after the material cools to its transition temperature.

Thus, if a magnetic field is applied after the perfect conductor is cooled, that field will not be able to penetrate that material (because the material's interior had no magnetic field prior to cooling). The top half of **Figure 9** illustrates the magnetic field's behavior when a perfect conductor is cooled before the magnetic field is applied. Additionally, if a magnetic field is present in a perfect conductor, it will remain in that material after it is cooled to its transition temperature (since the magnetic flux cannot change spatially, it will remain in the material). The bottom half of **Figure 9** illustrates the magnetic field's behavior when a perfect conductor is cooled after the magnetic field is applied. Therefore, this theoretical model can explain the response of perfect conductors to applied magnetic fields.

9. Permittivity examination

Superconductors exhibit the properties of zero electrical resistance as well as magnetic field exclusion because the material's permittivity becomes infinite. It can be shown that the permittivity of atoms can become infinite under specified conditions [27]. Atoms possess a resonant frequency and if they are excited at this frequency, their permittivity becomes infinitely large. So, if or when the frequency of the conduction electrons aligns with the resonant frequency of the atoms, sufficient conditions will be met (and the permittivity of the material will become infinite).

In general, a graph of a material's permittivity as a function of frequency displays three notable regions. In the microwave region or below 10^9 Hz the permittivity is constant, in the infrared region or at approximately 10^{12} Hz the permittivity displays a spike or discontinuity, and in the ultraviolet region or at approximately 10^{15} Hz the permittivity displays a second spike or discontinuity. A typical graph of this response is shown in **Figure 10** [32].

These spikes or discontinuities reveal that the permittivity goes to infinity at these frequencies. Therefore, these types of graphs alone indicate that the permittivity can be infinitely large and thus according to the theoretical model, the potential energy vanishes in the Schrödinger wave equation. When the potential energy term becomes zero, conduction electrons experience no impedance and thus superconductivity results. So, if conduction electrons in a material exist with a frequency that matches the frequency at one of these permittivity spikes, then the necessary conditions will exist that lead to superconductivity.

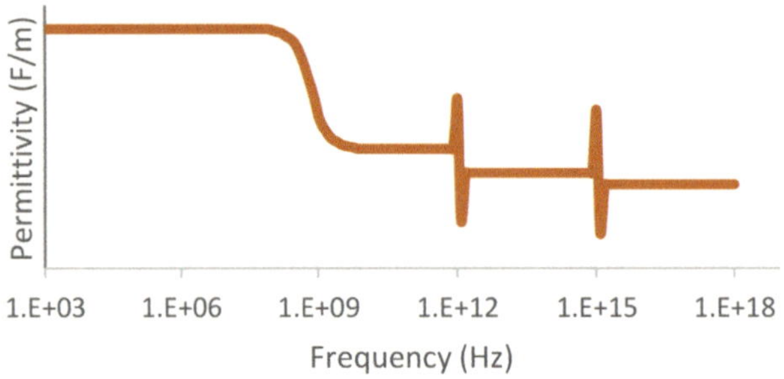

Figure 10.
Illustration of the permittivity as a function of frequency for a generic material. These first discontinuity, which occurs around 10^{12} Hz represents the resonance that occurs for atom-atom interactions. The second discontinuity, which occurs around 10^{15} Hz represents the resonance that occurs for electron-atom interactions.

The frequency of conduction electrons can be determined by using the Planck-Einstein relationship between energy and frequency. In equation form, $E = hf$, where E is the energy of the electron, h is Planck's constant, and f is the frequency of the electron. It is understood that conduction electrons have larger energies compared to other electrons in a material. These energies are approximately equal to the Fermi energy and will depend upon the material. Fermi energies have been determined for many materials and can range from approximately 1 to 10 eV. Using an average energy value of 5 eV, this corresponds to a frequency of 10^{15} Hz [1].

Comparing conduction electron frequencies to the discontinuities in the permittivity spectrum, it is seen that these two frequencies can align at 10^{15} Hz. Depending upon a material's characteristics, these two frequencies will align for some materials but will not align for others. This general analysis confirms that the mechanisms and analysis associated with the theoretical model are plausible and thus will lead to the phenomena of superconductivity.

10. Summary and discussion

A theoretical model has been developed that explains why some materials behave like superconductors (and thus display the corresponding electrical and magnetic properties), and why other materials do not. The theoretical model produces results in which electrical resistance is a function of temperature as well as results that explain why magnetic fields can or cannot exist in these materials. These theoretical results are validated by experimental results regardless of whether a material is classified as a type I superconductor, a type II superconductor, or a non-superconductor.

It is reasonable to state that atoms have a temperature dependent frequency response since energy, temperature, and an atom's motion are directly related [33]. The theoretical model and material properties can be used to help design superconducting materials that achieve zero resistance at specified temperatures. Having knowledge of how atoms function in materials will enhance the ability of scientists and engineers to create materials with specific properties. Therefore, by understanding the temperature and frequency relationship of atoms, engineers can manufacture or create materials that operate with zero resistance under desired conditions [34]. Moreover, since type II superconductors are better than type I materials at withstanding higher magnetic fields (before they exhibit non-superconducting properties), engineers can also use this knowledge of atomic behavior in materials to create superconductors with desired properties.

The electrical resistance for superconducting materials is linear for temperatures above the critical temperature, zero for temperatures below the critical temperature, and has a transition region at the transition temperature. For type I materials, the transition occurs within a narrow temperature range (approximately 10^{-3} K), whereas for type II materials the transition occurs over a large temperature range (approximately 1 K) [28]. The value and slope of the transition to zero resistance can be affected by many factors such as purity and the presence of isotopes [28, 35, 36]. Because the electrical resistance response of type II materials is prone to these lattice irregularities, they can be considered as type I materials that are either impure materials or composite materials. Therefore, it is reasonable to alter the theoretical model for type I materials and modify the characteristics of the atoms to obtain the response for type II materials.

It has been determined that permittivity is the electromagnetic parameter that is of importance in determining whether a material is a superconductor or not. Because permittivity is the governing parameter and not conductivity, ultimately it would probably be better to model the superconducting channel as a capacitor rather than a resistance. These two electrical circuit elements are similar in that they will limit electrical current. However, a capacitor would offer impedance rather than resistance and thus frequency effects could be modeled [37, 38]. Nevertheless, since impedance and resistance are similar concepts, if frequency effects are not important to an application (and in many applications, frequency effects are not important), the main characteristics associated with superconductors can still be modeled quite well using the resistor concept.

Sometimes a classical physics approach to modeling a system can provide accurate results and insight into the behavior of that system [39–41]. When a classical approach is used to analyze atoms in a material, it can be shown that the atoms have a resonant frequency. Furthermore, it can be shown that the atoms have a permittivity that is dependent upon this resonant frequency [27]. If the frequency of conduction electrons matches the resonant frequency of the atoms, the permittivity will become infinitely large and the material will behave as a superconductor.

Based on this resonant frequency analysis, at sufficiently low temperatures, superconducting atoms will have a resonant frequency that aligns with the frequency of conduction electrons. This allows conduction electrons to travel through the material without encountering electrical potentials. However, the conduction electron and atom frequency alignment will not occur at elevated temperatures due to effects of thermal energy on atoms. Conversely, reducing the temperature of non-superconducting materials will not lead to superconductivity because the atoms (or molecules) in non-superconducting materials have a resonant frequency that will not match the frequency of the conduction electrons.

In applying atomic theory to the resistance analysis, note that the resonant frequency of the atoms is normally 'hidden' at elevated temperatures. An additional frequency will be superimposed onto the resonant frequency of the atoms due to thermal energy. As a result, the resonant frequency will be masked if the temperature is above the critical temperature [42–44]. When the temperature of the superconductor is lowered to the threshold or critical value, the atoms in the material will display their resonance and therefore the material will behave as a superconductor.

In addition to verifying the electrical properties, this theoretical model also accounts for the magnetic field properties associated with superconductors. In other words, the model explains why the Meissner effect occurs. A material's permittivity and permeability are linked through Ampere's law. So, for a material that behaves as a superconductor, when its permittivity becomes infinitely large at sufficiently low temperatures, its permeability must go to zero. Therefore, the material must exclude magnetic fields from its interior. Surface currents produce magnetic fields and will account for this cancelation.

This theoretical model can also explain why the mixed state occurs in type II superconductors (i.e., the model explains why the magnetic field can exist in specified regions of the material). A magnetic field can penetrate the material because when a strong field is applied, the atoms are stretched and distorted and as a result the permittivity will no longer be infinite [45]. Perpetual currents will not be able to exist in the interior of the material to shield the external magnetic field. However, surface currents that exist above the atoms will be able to shield some of the external magnetic field. A fraction of the magnetic field will penetrate this surface shield and will exist inside the material.

To further explain the mixed state or why the magnetic field phenomena occurs in type II materials, consider the following. Superconductors have well known relationships between temperature and resistivity as well as between temperature and magnetic field strength. Because of these relationships, when a magnetic field is applied to a superconductor, that field will have an effect on the electrical resistance [46–48]. Since the magnetic field will alter the relationship between the atoms and electrons, if the field becomes large enough, it will alter the resonant frequency of the atoms. As a result, the permittivity ε will no longer be infinite, but it will become finite. Thus, the atoms in the bulk of the material will exhibit electrical resistance (and electrical currents will not flow indefinitely) and the atoms will not have the ability to block external magnetic fields.

11. Conclusion

A theoretical model has been developed based on atomic level analysis and electromagnetic field theory to explain why some materials exhibit superconductor properties and other materials do not. Specifically, the theoretical model addresses electrical resistance and the material's response to magnetic fields. The model explains why the electrical resistance of superconductors becomes zero or why conduction electrons are unimpeded by atoms. The model also explains why superconductors exhibit the Meissner effect and exclude magnetic fields from their interior. Additionally, the theoretical model is general enough such that it can describe the behavior of all materials (whether they are superconductors or not), but the model is specific enough such that it can explain the behavior of both type I or type II superconductors. Furthermore, the theoretical model also describes or distinguishes the characteristics of superconductors and perfect conductors and thus it is able to differentiate the behavior of these two materials.

Because many theories on superconductivity address specific aspects of superconductors and do not (or cannot) address other aspects, those theories have constraints and thus there are restrictions on the information they provide. Based on the approach and analysis that was used to construct this theoretical model (i.e., atomic physics and electromagnetic field theory), it provides insight into the physical mechanisms that cause materials to become superconducting. Therefore, this theoretical model should aid science and engineering researchers in the quest to develop room temperature superconductors that can be used to produce intense magnetic fields needed for medical imaging, as well as zero electrical resistance needed for supercomputing, and electric power transmission.

Conflict of interest

The author has no conflict of interest.

Author details

Fred Lacy
Electrical Engineering Department, Southern University and A&M College,
Baton Rouge, LA, USA

*Address all correspondence to: fred_lacy@subr.edu

References

[1] Kittel C. Introduction to Solid State Physics. Hoboken New Jersey: John Wiley and Sons; 2005. 704 p. DOI: 9780471415268.

[2] Buckel W, Kleiner R. Superconductivity (Fundamentals and Applications). Weinheim: Wiley-VCH; 2004. 655 p. DOI: 9783527618514

[3] Buzea C, Robbie K. Assembling the Puzzle of Superconducting Elements: A Review. Superconducting Science and Technology. 2004 Nov; 18(1): R1–8. DOI: 10.1088/0953-2048/18/1/R01

[4] Probst C, Wittig J. Chapter 10 Superconductivity: Metals, Alloys, and Compounds. Handbook on the Physics and Chemistry of Rare Earths. 1978; 1: 749–795. DOI: 10.1016/S0168-1273(78) 01014-4

[5] Marouchkine A. Room-Temperature Superconductivity, Cambridge UK, Cambridge International Science Publishing, 2004. 327 p. DOI: 1–904602–27-4

[6] *Ferry DK. Ohm's Law in a Quantum World. Science. 2012 Jan; 335(6064):45–46. DOI:* 10.1126/science.1215900

[7] *Meaden GT. Electrical Resistance of Metals. New York: Springer; 1965. 218 p. DOI:* 10.1007/978-1-4899-5717-7

[8] Hull JR, Murakami M. Applications of Bulk High-Temperature Superconductors. Proceedings of the IEEE. 2004 Oct; 92(10): 1705–1718. DOI: 10.1109/JPROC.2004.833796

[9] Malozemoff AP. Computer Applications of High Temperature Superconductivity. Physica C: Superconductivity. 1988 June; 153–155 (3): 1049–1054. DOI: 10.1016/ 0921-4534(88)90198-0.

[10] Durrell JH, Ainslie MD, Zhou D, Vanderbemden P, Bradshaw T, Speller S, Filipenko M, Cardwell DA. Bulk superconductors: a roadmap to applications. Superconductor science and technology. 2018; 31(10): 103501. DOI 10.1088/1361-6668/aad7ce

[11] Nakamura T, Itoh Y, Yoshikawa M, Oka T, Uzawa J. Development of a superconducting magnet for nuclear magnetic resonance using bulk high-temperature superconducting materials. Concepts in Magnetic Resonance Part B: Magnetic Resonance Engineering. 2007; 31(2): 65–70. DOI 10.1002/cmr.b.20083

[12] Lvovsky Y, Stautner EW, Zhang T. Novel technologies and configurations of superconducting magnets for MRI. Superconductor Science and Technology. 2013; 26(9): 093001. DOI 10.1088/0953-2048/26/9/093001

[13] Uglietti D. A review of commercial high temperature superconducting materials for large magnets: from wires and tapes to cables and conductors. Superconductor Science and Technology. 2019; 32(5): 053001. DOI 10.1088/1361-6668/ab06a2

[14] Kasap S, Capper P, editors. Handbook of Electronic and Photonic Materials. Springer; 2007. 1407 p. DOI: 978–0–387-29185-7

[15] File J, Mills RG. Observation of Persistent Current in a Superconducting Solenoid. Phys. Rev. Lett. 1963 Feb; 10 (3): 93–96. DOI: 10.1103/ PhysRevLett.10.93

[16] Schmalian J. Failed Theories of Superconductivity. Modern Physics Letters B. 2010; 24(27): 2679–2691. DOI: 10.1142/S0217984910025280

[17] Hirsch JE. Inconsistency of the Conventional Theory of Superconductivity. Europhysics Letters. 2020; 130(1): 17006–17012. DOI: 10.1209/0295-5075/130/17006

[18] Hirsch JE, BCS theory of superconductivity: it is time to question its validity, Physica Scripta 80(3), 2009. 035702.

[19] Zaanen J, Condensed-matter physics: Superconducting electrons go missing, Nature, 536(7616), p.282–3, Aug. 2016.

[20] Božović I, Bollinger AT, Wu J, He X, Can high-Tc superconductivity in cuprates be explained by the conventional BCS theory?, Low Temperature Physics., 44(6), p. 519–27, June 2018.

[21] Lacy F. Evaluating the Resistivity-Temperature Relationship for RTDs and other Conductors. IEEE Sensors Journal. 2011 May; 11(5): 1208–1213. DOI: 10.1109/JSEN.2010.2089977

[22] Lacy F. An Examination and Validation of the Theoretical Resistivity-Temperature Relationship for Conductors. International Journal of Electrical and Computer Engineering. 2013 Oct; 7(4): 439–443. DOI: 10.5281/zenodo.1335836

[23] da Luz MS, Dos Santos CA, Sandim MJ, Machado AJ, Jardim RF. Transport properties of granular high-T C superconductors. Brazilian Journal of Physics. 2007 Sep; 37(3B): 1155–9.

[24] Rahman MA, Rahaman MZ, Samsuddoha MN. A review on cuprate based superconducting materials including characteristics and applications. American Journal of Physics and Applications. 2015 Mar; 3(2): 39–56.

[25] Lacy F. Electrical Resistance of Superconductors. In: Proceedings of the World Congress on Engineering and Computer Science (WCECS 2019); 22–24 Oct 2019; San Francisco. 2019. p. 1–5.

[26] Lacy F. Understanding the Behavior of Superconductors by Analyzing Permittivity. International Journal of Electrical and Computer Engineering. 2021 Jan; 15(2):60–7.

[27] Balanis C. Advanced Engineering Electromagnetics. Hoboken New Jersey: John Wiley and Sons; 1989. 1008 p. DOI: 9780471621942

[28] Wesche R. High-Temperature Superconductors. In: Kasap S, Capper P, editors. Handbook of Electronic and Photonic Materials. Springer; 2017. p. 1225–1256. DOI: 10.1007/978-3-319-48933-9_50

[29] London F, London H. The electromagnetic equations of the supraconductor. Proc. R. Soc. A. 1935; 149: 71–88. DOI: 10.1098/rspa.1935.0048

[30] Abrikosov AA. On the magnetic properties of second kind superconductors. Sov. Phys. JETP. 1957; 5(6): 1174–1182.

[31] Henyey FS. Distinction between a Perfect Conductor and a Superconductor. Phys. Rev. Lett. 1982; 49(6): 416. DOI: 10.1103/PhysRevLett.49.416

[32] Yang X, Liu X, Yu S, Gan L, Zhou J, Zeng Y. Permittivity of Undoped Silicon in the Millimeter Wave Range. Electronics. 2019; 8(8): 886. DOI: 10.3390/electronics8080886

[33] Zhang S, Sahin H, Torun E, Peeters F, Martien D, DaPron T, Dilley N, Newman N. Fundamental mechanisms responsible for the temperature coefficient of resonant frequency in microwave dielectric ceramics. Journal of the American Ceramic Society. 2017; 100(4): 1508–1516. DOI: 10.111/jace.14648.

[34] Wu L, Xi X, Li B, Zhou J. Dielectric meta-atom with tunable resonant frequency temperature coefficient. Scientific Reports. 2017 May; 7(1): 1–6. DOI: 10.1038/s41598-017-02974-9

[35] Geballe TH, Matthias BT. Isotope effects in low temperature superconductors. IBM Journal of Research and Development. 1962; 6(2): 256–257. DOI: 10.1147/rd.62.0256

[36] Budnick JI. Some studies of the superconducting transition in purified tantalum. Physical Review. 1960; 119 (5): 1578–1586. DOI: 10.1103/ PhysRev.119.1578

[37] G. Ciovati. AC/RF superconductivity. CERN, Geneva, Switzerland, CERN Yellow Rep. CERN-2014-005, 57–75, Jan 2015. DOI: : 10.5170/CERN-2014-005.57

[38] Pompeo N, Torokhtii K, Silva E. Dielectric resonators for the measurements of the surface impedance of superconducting films. Measurement Science Review. 2014 Jun 1; 14(3): 164–70. DOI:10.2478/msr-2014-0022

[39] Hashimoto Y, Seniutinas G, Balčytis A, Juodkazis S, Nishijima Y. Au-Ag-Cu nano-alloys: tailoring of permittivity. Scientific reports. 2016 Apr 27;6:25010.

[40] Kadi M, Smaali A, Outemzabet R. Analysis of optical and related properties of tin oxide thin films determined by Drude-Lorentz model. Surface and Coatings Technology. 2012 Oct 25;211:45–9.

[41] Romanyuk V, Dmitruk N, Karpyna V, Lashkarev G, Popovych V, Dranchuk M, Pietruszka R, Godlewski M, Dovbeshko G, Timofeeva I, Kondratenko O. Optical and electrical properties of highly doped ZnO: Al films deposited by atomic layer deposition on Si substrates in visible and near infrared region. Acta Phys. Pol. A. 2016 Jan 1;129(1):A36–40.

[42] Nyquist H. Thermal agitation of electric charge in conductors. Phys. Rev. 1928; 32: 110–113. DOI: 10.1103/ PhysRev.32.110

[43] Raman CV. The Vibration Spectrum of a Crystal Lattice. Proc. Ind. Acad. Sci. 1943; A18: 237–250. DOI: 10.1007/ BF03046531.

[44] Butt HJ, Jaschke M. Calculation of Thermal Noise in Atomic Force Microscopy. Nanotechnology. 1995; 6 (1): 1–7; DOI: 10.1088/0957-4484/6/1/ 001.

[45] Krems, RV. Molecules near absolute zero and external field control of atomic and molecular dynamics. International Reviews in Physical Chemistry. 2005; 24 (1): 99–118. DOI: 10.1080/ 01442350500167161

[46] Bricefio G, Crommie MF, Zettl A. Out-of-plane current transport in Bi2Sr2CaCu208 in the mixed state. Physica C: Superconductivity. 1993 Jan; 204(3): 389–393. DOI:10.1016/ 0921-4534(93)91024-P

[47] Zuo F, Zuo F, Schlueter JA, Kelly ME, Williams JM. Mixed-state magnetoresistance in organic superconductors k–(BEDT-TTF)2Cu (NCS)2. Phys. Rev. B. 1996 Nov; 54(17): 11973–11976. DOI: 10.1103/ PhysRevB.54.11973

[48] Wu MK, Ashburn JR, Torng CJ. Superconductivity at 93 K in a New Mixed-Phase Y-Ba-Cu-O Compound System at Ambient Pressure. Phys. Rev. Letters. 1987 Mar; 58(9), 908–910. DOI: 10.1103/PhysRevLett.58.908

Chapter 5

Physics of Absorption and Generation of Electromagnetic Radiation

Sukhmander Singh, Ashish Tyagi and Bhavna Vidhani

Abstract

The chapter is divided into two parts. In the first part, the chapter discusses the theory of propagation of electromagnetic waves in different media with the help of Maxwell's equations of electromagnetic fields. The electromagnetic waves with low frequency are suitable for the communication in sea water and are illustrated with numerical examples. The underwater communication have been used for the oil (gas) field monitoring, underwater vehicles, coastline protection, oceanographic data collection, etc. The mathematical expression of penetration depth of electromagnetic waves is derived. The significance of penetration depth (skin depth) and loss angle are clarified with numerical examples. The interaction of electromagnetic waves with human tissue is also discussed. When an electric field is applied to a dielectric, the material takes a finite amount of time to polarize. The imaginary part of the permittivity is corresponds to the absorption length of radiation inside biological tissue. In the second part of the chapter, it has been shown that a high frequency wave can be generated through plasma under the presence of electron beam. The electron beam affects the oscillations of plasma and triggers the instability called as electron beam instability. In this section, we use magnetohydrodynamics theory to obtain the modified dispersion relation under the presence of electron beam with the help of the Poisson's equation. The high frequency instability in plasma grow with the magnetic field, wave length, collision frequency and the beam density. The growth rate linearly increases with collision frequency of electrons but it is decreases with the drift velocity of electrons. The real frequency of the instability increases with magnetic field, azimuthal wave number and beam density. The real frequency is almost independent with the collision frequency of the electrons.

Keywords: electromagnetic waves, permittivity, skin depth, loss angle, absorption, Dispersion equations, electron collisions, growth rate, Hall thruster, beam, resistive instability

1. Introduction

X-rays are used to detect bone fracture and determine the crystals structure. The electromagnetic radiation are also used to guide airplanes and missile systems. Gamma rays are used in radio therapy for the treatment of cancer and tumor Gamma rays are used to produce nuclear reaction. The earth get heat from Infrared waves. It is used to kill microorganism. Ultraviolet rays are used for the sterilizing of

The Electromagnetic Spectrum				
Frequency (Hz)	Nature	Wavelength (m)	Production	Applications
10^{22}	gamma rays	10^{-13}	Nuclear decay	Cosmic rays
10^{21}	gamma rays	10^{-12}	Nuclear decay	Cancer therapy
10^{18}	x rays	10^{-9}	Inner electronic transitions and fast collisions	Medical diagnosis
10^{16}	ultraviolet	10^{-7}		Sterilization
10^{15}	visible	10^{-6}	Thermal agitation and electronic transitions	Vision, astronomy, optical
6.5×10^{14}	blue	4.6×10^{-7}		
5.6×10^{14}	green	5.4×10^{-7}		
3.9×10^{14}	red	7.6×10^{-7}		
10^{14}	infrared	10^{-5}	Thermal agitation and electronic transitions	Heating, night vision, optical communications
10^{9}	UHF	10^{-3}	Accelerating charges and thermal agitation	Microwave ovens
10^{10}	EHF	10^{-1}		Remote sensing
10^{8}	TV FM	10		Radio transmission
10^{6}	AM	10^{3}		Radio signals
10^{4}	RF	10^{5}	Accelerating charges	

Table 1.
The electromagnetic spectrum.

surgical instruments. It is also used for study molecular structure and in high resolving power microscope. The color of an object is due to the reflection or transmission of different colors of light. For example, a fire truck appears red because it reflects red light and absorbs more green and blue wavelengths. Electromagnetic waves have a huge range of applications in broadcasting, WiFi, cooking, vision, medical imaging, and treating cancer. Sequential arrangement of electromagnetic waves according to their frequencies or wave lengths in the form of distinct of groups having different properties in called electromagnetic spectrum. In this section, we discuss how electromagnetic waves are classified into categories such as radio, infrared, ultraviolet, which are classified in **Table 1**. We also summarize some of the main applications for each range of electromagnetic waves. Radio waves are commonly used for audio communications with wavelengths greater than about 0.1 m. Radio waves are produced from an alternating current flowing in an antenna.

2. Current status of the research

Underwater communications have been performed by acoustic and optical systems. But the performance of underwater communications is affected by multipath propagation in the shallow water. The optical systems have higher propagation speed than underwater acoustic waves but the strong backscattering due to suspended particles in water always limits the performance of optical systems [1]. UV radiation, free radicals and shock waves generated from electromagnetic fields are effectively used to sterilize bacteria. Pulsed electromagnetic fields (streamer discharge) in water are employed for the sterilization of bacteria. For biological applications of pulsed electromagnetic field, electroporation is usually used to sterilize bacteria. This technique is commonly applied for sterilization in food processing. The cells in the region of tissue hit by the laser beam (high intensities $\sim 10 - 100\ W/cm^2$) usually dies and the resulting region of tissue burn is called a photocoagulation burn. Photocoagulation burns are used to destroy tumors, treat eye conditions and stop bleeding.

Electromagnetic waves in the RF range can also be used for underwater wireless communication systems. The velocity of EM waves in water is more than 4 orders faster than acoustic waves so the channel latency is greatly reduced. In addition, EM waves are less sensitive than acoustic waves to reflection and refraction effects in shallow water. Moreover, suspended particles have very little impact on EM waves. Few underwater communication systems (based on EM waves) have been proposed in reference [2, 3]. The primary limitation of EM wave propagation in water is the high attenuation due to the conductivity of water. For example, it has been shown in [4] that conventional RF propagation works poorly in seawater due to the losses caused by the high conductivity of seawater (typically, 4 S/m). However, fresh water has a typical conductivity of only 0.01 S/m, which is 400 times less than the typical conductivity of seawater. Therefore, EM wave propagation can be more efficient in fresh water than in seawater. Jiang, and Georgakopoulos analyzed the propagation and transmission losses for a plane wave propagating from air to water (frequency range of 23 kHz to 1 GHz). It has been depicted that the propagation loss increases as the depth increases, whereas the transmission loss remains the same for all propagation depths [5]. Mazharimousavi et al. considered variable permeability and permittivity to solve the wave equation in material layers [6]. The Compton and Raman scattering effects are widely employed in the concept of free electron lasers. These nonlinear effects have great importance for fusion physics, laser-plasma acceleration and EM-field harmonic generation. Matsko and Rostovtsev investigated the behavior of overdense plasmas in the presence of the Electromagnetic fields, which can lead to the nonlinear effects such as Raman scattering, modulational instability and self-focusing [7]. The increasing relativistic mass of the particles can make plasma transparent in the presence of high intense the electromagnetic field change the properties of plasmas [8]. The models of electromagnetic field generated in a non absorbing anisotropic multilayer used to study the optical properties of liquid crystals and propagation of electromagnetic waves in magneto active plasmas [9]. Pulse power generator based on electromagnetic theory has applications such as water treatment, ozone generation, food processing, exhaust gas treatment, engine ignition, medical treatment and ion implantation. The similar work was reviewed by Akiyama et al. [10]

Applications for environmental fields involving the decomposition of harmful gases, generation of ozone, and water treatment by discharge plasmas in water utilizing pulsed power discharges have been studied [11–14]. High power microwave can be involved to joining of solid materials, to heat a surface of dielectric material and synthesis of nanocomposite powders. Bruce et al. used a high-power millimeter wave beam for joining ceramics tubes with the help of 83-GHz Gyrotron [15]. The use of shock waves to break up urinary calculi without surgery, is called as extracorporeal shock wave lithotripsy. Biofilm removal to inactivation of fungi, gene therapy and oncology are the interesting uses of shock waves lithotripsy. Loske overviewed the biomedical applications (orthopedics, cardiology, traumatology, rehabilitation, esthetic therapy) of shock waves including some current research. [16]. Watts et al. have reported the theory, characterization and fabrications of metamaterial perfect absorbers (MPAs) of electromagnetic waves. The motivation for studying MPAs comes mainly from their use in potential applications as selective thermal emitters in automotive radar, in local area wireless network at the frequency range of 92–95 GHz and in imaging at frequency 95 and 110 GHz. [17]. Ayala investigated the applications of millimeter waves for radar sensors [18]. Metamaterial perfect absorbers are useful for spectroscopy and imaging, actively integrated photonic circuits and microwave-to-infrared signature control [19–21]. In [22, 23], authors show the importance of THz pulse imaging system for characterizing biological tissues such as skin, muscle and veins. Reference [24] reported the propagation of EM waves on a graphene sheet. The Reference [25] compared the CNT-based nano dipole antenna and GNR-based nano patch antenna. Due to short wavelength, even a minute variations in water contents and biomaterial tissues can be detected by terahertz radiations due to existence of molecular resonances at such frequencies. Consequently, one of the emerging areas of research is analyzing the propagation of terahertz electromagnetic waves through the tissues to develop diagnostic tools for early detection and treatment such as abnormalities in skin tissues as a sign of skin cancer [26]. Shock waves may stimulate osteogenesis and chondrogenesis effects [27], induce analgesic effects [28] and tissue repair mechanisms [29]. Shock waves therapy are also used to treat oncological diseases and other hereditary disorders [27, 30]. Chen et al. proposed a mathematical model for the propagating of electromagnetic waves coupling for deep implants and simulated through COMSOL Multiphysics [31]. Body area networks technological is used to monitor medical sensors implanted or worn on the body, which measure important physical and physiological parameters [32, 33]. Marani and Perri reviewed the aspects of Radio Frequency Identification technology for the realization of miniaturized devices, which are implantable in the human body [34]. Ultrasonic can transport high power and can penetrate to a deeper tissue with better power efficiency. [35, 36]. Ref. [37], discuss the radar-based techniques to detect human motions, wireless implantable devices and the characterization of biological materials. Low frequency can deliver more power with deeper penetrating ability in tissue [38, 39]. Contactless imaging techniques based on electromagnetic waves are under continuous research. Magnetic resonance imaging technology and physiological processes of biological tissues and organisms [40, 41]. The electrical properties of biological mediums are found very useful because it is related to the pathological and physiological state of the tissues [42–44].

3. Interaction of electromagnetic wave fields with biological tissues

From last decade, researchers are interested about biological effects of electromagnetic energy due to public concern with radiation safety and measures. The electromagnetic energy produces heating effects in the biological tissues by increasing the kinetic energy of the absorbing molecules. Therefore the body tissues absorb strongly in the UV and in the Blue/green portion of the spectrum and transmit reds and IR. A surgeon can select a particular laser to target cells for photovaporization by determining which wavelengths your damaged cell will absorb and what the surrounding tissue will not. The heating of biological tissues depends on dielectric properties of the tissues, tissue geometry and frequency of the source. The tissues of the human body are extremely complex. Biological tissues are composed of the extracellular matrix (ECM), cells and the signaling systems. The signaling systems are encoded by genes in the nuclei of the cells. The cells in the tissues reside in a complex extracellular matrix environment of proteins, carbohydrates and intracellular fluid composed of several salt ions, polar water molecules and polar protein molecules. The dielectric constant of tissues decreases as the frequency is increased to GHz level. The effective conductivity, rises with frequency. The tissues of brain, muscle, liver, kidney and heart have larger dielectric constant and conductivity as compared to tissues of fat, bone and lung. The action of electromagnetic fields on the tissues produce the rotation of dipole molecules at the frequency of the applied electromagnetic energy which in turn affects the displacement current through the medium with an associated dielectric loss due to viscosity. The electromagnetic field also produce the oscillation of the free charges, which in turn gives rise to conduction currents with an associated energy loss due to electrical resistance of the medium. The interaction of electromagnetic wave fields with biological tissues is related to dielectric properties. Johnson and Guy reviewed the absorption and scattering effects of light in biological tissues [45]. In ref. [46], the method of warming of human blood from refrigerated (bank blood storage temperature $\sim$ 4 to 6°C) has been discussed with the help of microwave.

4. Complex dielectric permittivity

The dielectric permittivity of a material is a complex number containing both real and imaginary components. It describes a material's ability to permit an electric field. It dependent on the frequency, temperature and the properties of the material. This can be expressed by

$$\varepsilon_c = \varepsilon_0(\varepsilon' - j\varepsilon'') \tag{1}$$

where ε' is the dielectric constant of the medium. The ε'' is called the loss factor of the medium and related with the effective conductivity such that $\varepsilon'' = \frac{\sigma}{\varepsilon_0 \omega}$. These coefficients are related through by loss tangent $\tan\delta = \frac{\varepsilon''}{\varepsilon'}$. In other words loss factor is the product of loss tangent and dielectric constant, that is $\varepsilon'' = \varepsilon' \tan\delta$. The loss tangent depends on frequency, moisture content and temperature. If all energy is dissipated and there is no charging current then the loss tangent would tend to infinity and if no energy is dissipated, the loss tangent is zero [45, 47–49]. The high power electromagnetic waves are used to generate plasma through laser plasma interaction. Gaseous particles are ionized to bring it in the form of plasma through injection of high frequency microwaves. The electrical permittivity in plasma is affected by the plasma density [50]. If the microwave electric field ($\tilde{E}$) and the velocity ($\tilde{v}$) are assumed to be varying with $e^{i\omega t}$, the plasma dielectric constant can be read as,

$$\epsilon = \epsilon_0 \left(1 - \frac{\omega_{pe}^2}{\omega^2}\right) \tag{2}$$

Where; the ω_{pe} is the electron plasma frequency and given by the relation,

$$\omega_{pe} = \sqrt{\frac{n_e e^2}{\epsilon_0 m_e}} \tag{3}$$

Recently many researchers have studied the plasma instabilities in a crossed field devices called Hall thrusters (space propulsion technology). The dispersion relations for the low and high frequency electrostatic and electromagnetic waves are derived in the magnetized plasma. The dispersion relations for the resistive and Rayliegh Taylor instabilities has been derived for the propagation of waves in a magnetized plasma under the effects of various parameters [51–61].

5. Propagation of EM fields (waves) in conductors

The behavior of EM waves in a conductor is quite different from that in a source-free medium. The conduction current in a conductor is the cause of the difference. We shall analyze the source terms in the Maxwell's equations to simplify Maxwell's equations in a conductor. From this set of equations, we can derive a diffusion equation and investigate the skin effects.

5.1 Gauss' law for electric field

The Electric flux φ_E through a closed surface A is proportional to the net charge q enclosed within that surface.

$$\varphi_E = \oint \vec{E} \cdot \hat{n} dA = \frac{q}{\varepsilon_0} = \frac{1}{\varepsilon_0} \int_V \rho dV \tag{4}$$

$$\text{Differential form, } \vec{\nabla} \cdot \vec{E} = \frac{\rho}{\varepsilon_0} \tag{5}$$

5.2 Faraday's law

The electromagnetic force induced in a closed loop, is proportional to the negative of the rate of change of the magnetic flux, φ_B through the closed loop,

$$\oint \vec{E} \cdot d\vec{l} = \frac{\partial \varphi_B}{\partial t} = \frac{\partial}{\partial t} \oint \vec{B} \cdot dA \tag{6}$$

Faraday's law in differential form,

$$\vec{\nabla} \times \vec{E} = -\frac{\partial \vec{B}}{\partial t} \tag{7}$$

5.3 Magnetic Gauss's law for magnetic field

The Magnetic flux φ_B through a closed surface, A is equal to zero.

$$\varphi_B = \oint \vec{B} \cdot dA = 0 \tag{8}$$

In the differential form

$$\vec{\nabla} \cdot \vec{B} = 0 \tag{9}$$

5.4 Ampere's law

The path integral of the magnetic field around any closed loop, is proportional to the current enclosed by the loop plus the displacement current enclosed by the loop.

$$\oint \vec{B} \cdot d\vec{l} = \mu_0 I + \mu_0 \varepsilon_0 \frac{\partial \varphi_E}{\partial t} \tag{10}$$

Ampere's law in differential form

$$\vec{\nabla} \times \vec{B} = \mu \sigma \vec{E} + \mu \varepsilon \frac{\partial \vec{E}}{\partial t} \tag{11}$$

6. Properties of plane wave (monochromatic) in vacuum

Let us assume that the wave equations (fields) has the solution in the form of $\vec{E}\left(\vec{B}\right) = \vec{E}_0\left(\vec{B}_0\right)e^{-i(kz-\omega t)}$, then the vector operators can be written as $\nabla \to -ik$ and $\frac{\partial}{\partial t} \to i\omega$.

a. The vector k and fields $\vec{E}\left(\vec{B}\right)$ are perpendicular

From Gauss's law $k \cdot E = 0$

b. The field $\vec{B}$ is perpendicular to the vector k and field $\vec{E}$

From Faraday's law $-i\vec{k} \times \vec{E} = -i\omega\vec{B}$

$$\Rightarrow \vec{B} = \frac{\vec{k} \times \vec{E}}{\omega} = \frac{k\hat{k} \times \vec{E}}{\omega} = \frac{\hat{k} \times \vec{E}}{c} \tag{12}$$

Where we have used $\omega = ck$ and unit vector $\hat{k} = \vec{k}/_k$. This implies that all three vectors are perpendicular to one another (**Figure 1**).

Let us apply curl operator to the 2nd equation.

Maxwell's equation:

$$\vec{\nabla} \times \left(\vec{\nabla} \times \vec{E}\right) = -\vec{\nabla} \times \left(\frac{\partial \vec{B}}{\partial t}\right) = -\frac{\partial}{\partial t}\left(\vec{\nabla} \times \vec{B}\right) = -\frac{\partial}{\partial t}\left(\mu\sigma\vec{E} + \mu\varepsilon\frac{\partial \vec{E}}{\partial t}\right) \tag{13}$$

So

$$-\nabla^2\vec{E} = -\mu\sigma\frac{\partial \vec{E}}{\partial t} - \mu\varepsilon\frac{\partial^2 \vec{E}}{\partial t^2} \tag{14}$$

Similar, the magnetic field satisfy the same equation

$$-\nabla^2\vec{B} = -\mu\sigma\frac{\partial \vec{B}}{\partial t} - \mu\varepsilon\frac{\partial^2 \vec{B}}{\partial t^2} \tag{15}$$

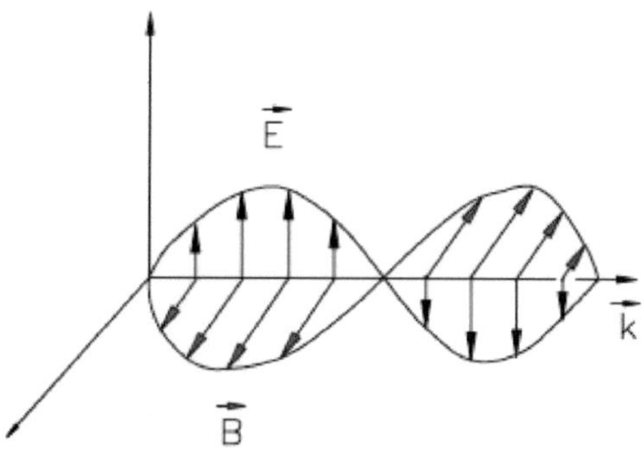

Figure 1.
Orientations of electric field, magnetic field and wave vector.

6.1 Skin depth

Suppose we have a plane wave field. It comes from the $-z$ direction and reaches a large conductor. Surface at $z = 0$ outside of a conductor: $E = E_0 e^{-i\omega t} e_x$ at $z = 0$. Looking for the wave like solution of electric (magnetic) fields by assuming the wave inside the conductor has the form, where k is an unknown constant. Suppose, the waves are traveling only in the z direction (no x or y components). These waves are called plane waves, because the fields are uniform over every plane perpendicular to the direction of propagation. We are interested, then, in fields of the form

$$\vec{E}\left(\vec{B}\right) = \vec{E}_0\left(\vec{B}_0\right)e^{-i(kz-\omega t)} \tag{16}$$

for the waves of the above type, we find from the diffusion equation

$$k^2 E = -i\omega\mu\sigma E + \mu\varepsilon\omega^2 E \tag{17}$$

Or $\left(k^2 + i\omega\mu\sigma - \mu\varepsilon\omega^2\right)E = 0$

$$\text{For non} - \text{trivial solution } k^2 + i\omega\mu\sigma - \mu\varepsilon\omega^2 = 0 \tag{18}$$

The presence of imaginary term due to conductivity of the medium gives different dispersion relation from the dielectric medium. From Eq. (18) we can expect the wave vector to have complex form.

Let us write

$$\vec{k} = \vec{\alpha} - i\vec{\beta} \tag{19}$$

Here the real part $\vec{\alpha}$ determine the wavelength, refractive index and the phase velocity of the wave in a conductor. The imaginary part $\vec{\beta}$ corresponds to the skin depth in a conductor. The solutions of Eqs. (18) and (19), gives the real and imaginary part of wave vector k in terms of materials' properties.

$$\alpha = \omega\sqrt{\frac{\varepsilon\mu}{2}}\left[\sqrt{1 + \frac{\sigma^2}{\varepsilon^2\omega^2}} + 1\right]^{\frac{1}{2}} \tag{20}$$

$$\text{And } \beta = \frac{\omega\mu\sigma}{2\alpha} \tag{21}$$

$$\text{Or } \beta = \omega\sqrt{\frac{\varepsilon\mu}{2}}\left[\sqrt{1 + \frac{\sigma^2}{\varepsilon^2\omega^2}} - 1\right]^{\frac{1}{2}} \tag{22}$$

If we use complex wave vector $\vec{k} = \vec{\alpha} - i\vec{\beta}$ into Eq. (16), then the wave equation for a conducting medium can be written as

$$\vec{E} = \vec{E}_0 e^{-\beta z} e^{-i(\alpha z - \omega t)} \tag{23}$$

It is clear from the above equation that the conductivity of the medium affects the wavelength for a fixed frequency. The first exponential factor $e^{-\beta z}$ gives an exponential decay in the amplitude (with increasing z) of the wave as shown in **Figure 2**. The cause of the decay of the amplitude of the wave can be explained in a very precise way in terms of conservation of energy. Whenever the incoming

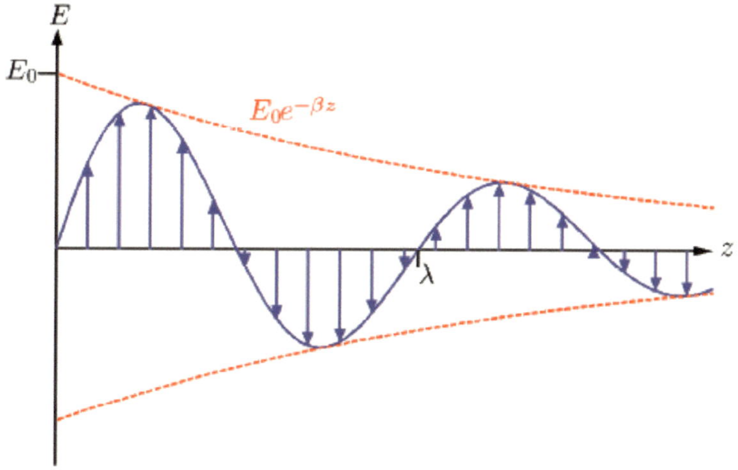

Figure 2.
Decaying of electromagnetic field.

electromagnetic radiation interacts with a conducting material, it produces current in the in the conductor. The current produces Joule heating effect which must be compensated from the energy of the wave. Therefore we can expect the decay in the amplitude of the wave. The second factor $e^{-i(\alpha z-\omega t)}$ gives the plane wave variations with space and time.

7. Alternating magnetic field in a conducting media

From Faraday's law, the both fields are related by

$$\vec{k} \times \vec{E}_0 = \omega \vec{B}_0 \tag{24}$$

$$\text{Or } \vec{B}_0 = \frac{\vec{k} \times \vec{E}_0}{\omega} \tag{25}$$

Thus as in dielectric case, both fields are perpendicular to each other and also perpendicular to the direction of motion with same phase angle.

7.1 Phase change in fields in a conducting media

The complex wave vector k, gives the phase angle between the fields in a conducting medium. Let us assume that E is polarized along the x direction

$$\vec{E} = \hat{i}\vec{E}_0 e^{-\beta z} e^{-i(\alpha z-\omega t)} \tag{26}$$

And the magnetic field results from Eq. (25) is given by

$$\vec{B} = \hat{j}\frac{|k|}{\omega}\vec{E}_0 e^{-\beta z} e^{-i(\alpha z-\omega t)} \tag{27}$$

From Eq. (19), the complex number k can be written as

$$|k| = \sqrt{\alpha^2 + \beta^2 e^{i\varphi}} = \text{Re}^{\,i\varphi} \quad (28)$$

$$\text{Thus } R = \sqrt{\alpha^2 + \beta^2} = \omega\sqrt{\varepsilon\mu}\left[\sqrt{1 + \frac{\sigma^2}{\varepsilon^2\omega^2}}\right]^{\frac{1}{4}} \quad (29)$$

$$\text{And the phase angle } \varphi = \tan^{-1}\frac{\beta}{\alpha} \quad (30)$$

Further if the initial phases of the fields are φ_E and φ_B, then the amplitude are given by

$$\vec{E}_0 = E_0 e^{i\varphi_E} \quad (31)$$

$$\vec{B}_0 = B_0 e^{i\varphi_B} \quad (32)$$

From Eq. (27)

$$B_0 e^{i\varphi_B} = \frac{\text{Re}^{\,i\varphi}}{\omega} E_0 e^{i\varphi_E} \quad (33)$$

Therefore, the both fields are out of phase with angle

$$\varphi = \varphi_B - \varphi_E \quad (34)$$

From Eq. (33) the ratio of the magnetic field to the electric field is

$$\frac{B_0}{E_0} = \frac{R}{\omega} \quad (35)$$

By using Eq. (29)

$$\frac{B_0}{E_0} = \sqrt{\varepsilon\mu}\left[1 + \frac{\sigma^2}{\varepsilon^2\omega^2}\right]^{\frac{1}{4}} \quad (36)$$

In other words, we can say that the magnetic field advanced from electric field by the phase angle φ. In terms of sinusoidal form, these fields follow the following expressions.

$$\vec{E} = \hat{i}\vec{E}_0 e^{-\beta z}\cos(\omega t - \alpha z + \varphi_E) \quad (37)$$

And the magnetic field results from Eq. (32) is given by

$$\vec{B} = \hat{j}\vec{B}_0 e^{-\beta z}\cos(\omega t - \alpha z + \varphi_E + \varphi) \quad (38)$$

The above equations direct that the amplitude of an electromagnetic wave propagating (through a conductor) decays exponentially on a characteristic length scale, d, that is known as the skin-depth [48].

7.2 Skin depth

Skin depth measure the distance that the wave travels before it's amplitude falls to 1/e of its original value [48]. From Eq. (37), the amplitude of the wave falls by a factor 1/e in a distance $z = \frac{1}{\beta}$. In other words it is a measure of how far the wave penetrates into the conductor. Mathematically skin depth is denoted by δ, therefore

$$\delta = \frac{1}{\beta} \tag{39}$$

If we study poor conductor, which satisfies the inequality $\sigma < < \varepsilon\omega$, then Eqs. (20) and (21) leads to

$$\alpha \approx \omega\sqrt{\varepsilon\mu} \tag{40}$$

$$\text{And we know that } \beta = \frac{\omega\mu\sigma}{2\alpha} \tag{41}$$

Substitute the value of α into Eq. (32), we get

$$\text{Or } \beta \approx \omega \frac{\mu\sigma}{2\omega\sqrt{\varepsilon\mu}} \tag{42}$$

$$\text{Or } \beta \approx \frac{\sigma}{2}\sqrt{\frac{\mu}{\varepsilon}} \tag{43}$$

$$\text{The phase velocity } V_{ph} = \frac{\omega}{\alpha} \approx \frac{1}{\sqrt{\varepsilon\mu}} \tag{44}$$

$$\text{Skin depth in poor conductor } \delta = \frac{1}{\beta} = \frac{2}{\sigma}\sqrt{\frac{\varepsilon}{\mu}} \tag{45}$$

So, it is independent from the frequency.

The Eq. (41) state that at higher frequency, the absorbing parameter lost its significance that is $\beta < < \alpha$. We can conclude that at higher frequency the wavelength does not decay very fast in a poor conductor. Moreover both the fields are also in same phase by the relation $\omega B_0 = \alpha E_0$. Also the phase velocity is independent from the frequency [47].

8. Wave propagation in perfect conductors

The transmission lines and communication systems are made up with silver, copper and aluminum. In most cases these conductors satisfies the inequality $\sigma > > \varepsilon\omega$, then Eqs. (20) and (21) leads to

$$\alpha \approx \sqrt{\frac{\omega\mu\sigma}{2}} \tag{46}$$

$$\text{And } \beta \approx \sqrt{\frac{\omega\mu\sigma}{2}} \tag{47}$$

$$\text{Therefore } \beta \approx \alpha \tag{48}$$

$$\text{Skin depth } \delta = \frac{1}{\beta} = \frac{1}{\alpha} = \sqrt{\frac{2}{\omega\mu\sigma}} \tag{49}$$

The wave decays significantly within one wavelength. Since $\delta \propto \sqrt{1/\omega\sigma}$, the deep penetration occurs, when the inequality$\sigma < < \varepsilon\omega$ is satisfied (at Low frequency in a Poor conductor).

9. Electromagnetic wave propagation into water

EM wave propagation can be more efficient in fresh water than in seawater. The radiofrequency wave propagation works poorly in seawater due to the losses caused by the high conductivity of seawater. The limitation of EM wave propagation in water is the high attenuation due to the conductivity of water (typically, 4 S/m), however fresh water has a conductivity of 0.01 S/m. These properties are used to construct underwater sensor network based on electromagnetic waves to trace out the natural resources buried underwater, where the conventional optical water sensors are difficult to utilize in an underwater environment due to backscatter and absorptions [47].

Example:

For sea water,

$\mu = \mu_0 = 4\pi \times 10^{-7} N/A^2, \varepsilon \cong 81\varepsilon_0$ and $\sigma \approx 5(\Omega.m)^{-1}$.

The skin depth in poor conductor

$$\delta = \frac{2}{\sigma}\sqrt{\frac{\varepsilon}{\mu}} = \frac{2}{\sigma}\sqrt{\frac{81\varepsilon_0}{\mu_0}} \tag{50}$$

$$= \frac{2\sqrt{81}}{\sigma Z} = \frac{18}{5 \times 377} \approx 0.96\text{cm}. \tag{51}$$

If the sea water satisfies the inequality $\sigma < < \varepsilon\omega$, of poor conductor, which require

$$f = \frac{\omega}{2\pi} > > \frac{\sigma}{2\pi\varepsilon} = 10^9 Hz \tag{52}$$

Therefore at $10^9 Hz$ or $\lambda < < 30\ cm$, sea water behave as poor conductor. On the other hand at the radio frequency range $f < < 10^9$ Hz, the inequality $\sigma > > \varepsilon\omega$, can be satisfied, the skin depth $\delta = \sqrt{2/(\omega\mu\sigma)}$ is quite short. To reach a depth δ = 10 m, for communication with submarines,

$$f = \frac{\omega}{2\pi} = \frac{1}{\pi\sigma\mu\delta^2} \approx 500\text{Hz} \tag{53}$$

The wavelength in the air is about

$$\lambda = \frac{c}{f} = \frac{3 \times 10^8}{500} = 600\text{km} \tag{54}$$

The skin depth at different frequency in sea water are 277 m at 1 Hz, 8.76 m at 1KHz, 0.277 m at 1 MHz and 0.015 at 1GHz if the conductivity of sea-water is taken to about $\sigma = 3/\Omega m$ and $\varepsilon_r = 80$. These effects leads to severe restrictions for radio communication with submerged submarines. To overcome this, the communication must be performed with extremely low frequency waves generated by gigantic antennas [47].

9.1 Short wave communications

At 60 km to 100 km height from the earth, ionosphere plasma has a typical density of $10^{13}/m^3$, which gives the plasma frequency of order 28 MHz. the waves below this frequency shows reflections from the layer of ionosphere to reach the receiver's end. The conductivity of the earth is 10^{-2} S/m, Earth behave as a good conductor, if the inequality $\sigma >> \varepsilon\omega$ is satisfied. In other word

$$f << \frac{\sigma}{2\pi\varepsilon} = 180 \tag{55}$$

MHz, therefore below 20 MHz, the earth is good conductor.

Example: skin depth at $f = 60$ Hz for copper.

$$\delta = \sqrt{\frac{2}{2\pi \times 60 \times 4\pi \times 10^{-7} \times 6 \times 10^{7}}} = 8\text{mm} \tag{56}$$

The frequency dependent skin-depth in Copper ($\sigma = 6.25 \times 10^{7}/\Omega m$) can be expressed as $d = \frac{6}{\sqrt{f(Hz)}}$ cm. It says that the skin-depth is about 6 cm at 1 Hz and it reduced to 2 mm at 1 kHz. In other words it conclude that an oscillating electromagnetic signal of high frequency, transmits along the surface of the wire or on narrow layer of thickness of the order the skin-depth in a conductor. In the visible region ($\omega \sim 10^{15}$/s) of the spectrum, the skin depth for metals is on the order of $10A^0$. The skin depth is related with wavelength λ (inside conductor) as

$$\lambda = \frac{2\pi}{\alpha} = 2\pi\sqrt{\frac{2}{\omega\mu\sigma}} \tag{57}$$

$$\text{The phase velocity } V_{ph} = \frac{\omega}{\alpha} = \frac{\omega\lambda}{2\pi} \approx \sqrt{\frac{2\omega}{\mu\sigma}} \tag{58}$$

Therefore for a very good conductor, the real and imaginary part of the wave vector attain the same values. In this case the amplitude of the wave decays very fast with frequency as compared to bad conductor. The phase velocity of the wave in a good conductor depends on the frequency of the electromagnetic light. Consequently, an electromagnetic wave cannot penetrate more than a few skin-depths into a conducting medium. The skin-depth is smaller at higher frequencies. This implies that high frequency waves penetrate a shorter distance into a conductor than low frequency waves.

Question: Find the skin depths for silver at a frequency of 10^{10} Hz.

$$\text{Skin depth } \delta = \sqrt{\frac{2}{\omega\mu\sigma}} \tag{59}$$

$$\delta = \sqrt{\frac{2}{2\pi \times 10^{10} \times 4\pi \times 10^{-7} \times 6.25 \times 10^{7}}} = 6.4 \times 10^{-4}\text{mm} \tag{60}$$

Therefore, in microwave experiment, the field do not penetrate much beyond .00064 mm, so no point it's coating making further thicker. There is no advantage to construct AC transmission lines using wires with a radius much larger than the skin depth because the current flows mainly in the outer part of the conductor.

Question: wavelength and propagation speed in copper for radio waves at 1 MHz. compare the corresponding values in air (or vacuum). $\mu_0 = 4\pi \times 10^{-7}$ H/m.

From Eq. (40),

$$\lambda_{Cu} = \frac{2\pi}{\alpha_{Cu}} \text{ and } \alpha_{Cu} \approx \sqrt{\frac{\omega\mu\sigma_{Cu}}{2}} \tag{61}$$

Therefore, $\lambda_{Cu} = 2\pi\sqrt{\frac{2}{\omega\mu\sigma_{Cu}}}$

$$\lambda_{Cu} = 2\pi\sqrt{\frac{2}{2\pi \times 10^6 \times 4\pi \times 10^{-7} \times 6.25 \times 10^7}} = 0.4\text{mm} \tag{62}$$

The propagating velocity in copper $V_{ph} = \frac{\omega}{\alpha} = \frac{\omega\lambda}{2\pi}$

$$V_{ph} = 0.4 \times 10^{-3} \times 10^6 = 400\text{m/s} \tag{63}$$

The above parameters are quite different in vacuum as follow

$$\lambda_{Vacuum} = \frac{c}{\nu} = \frac{3 \times 10^8}{10^6} = 300\text{m} \tag{64}$$

There is no advantage to construct AC transmission lines using wires with a radius much larger than the skin depth because the current flows mainly in the outer part of the conductor.

10. Complex permittivity of bread dough and depth of penetration

After baking for few minutes, the relative permittivity of bread dough at frequency 600 MHz is $\varepsilon_{cr} = 23.1 - j11.85$. Calculate the depth of penetration of microwave.

Solution: the loss tangent of bread dough is

$$\tan\delta = \frac{11.85}{23.1} = 0.513 \tag{65}$$

The depth of penetration is given as

$$d \approx \frac{c\sqrt{2}}{2\pi f \varepsilon_r' \sqrt{\left(\sqrt{1 + \tan\delta^2} - 1\right)}} \tag{66}$$

After substituting all the parameters, we get

$$d \approx \frac{\sqrt{2} \times 3 \times 10^8}{2\pi \times 600 \times 10^6} \frac{1}{23.1\sqrt{\left(\sqrt{1 + (0.513)^2} - 1\right)}} \approx 6.65\text{cm} \tag{67}$$

It is worthy to note that the depth of penetration decreases with frequency.

11. The AC and DC conduction in plasma

Let the collision frequency of electrons with ions and ω the frequency of the EM waves in the conductor. The equation of motion for electrons is:

$$m\frac{dv}{dt} = -eE - m\nu v \tag{68}$$

Assume $v = v_0 e^{-i\omega t}$ and use $\partial/\partial t \to -i\omega$, we obtain

$$-i\omega m v = -eE - m\nu v \to v = \frac{-e}{m(\nu - i\omega)}E \tag{69}$$

the current density is expressed by $j = -env$

$$j_f = \frac{-ne^2}{m(\nu - i\omega)}E \tag{70}$$

Therefore, the AC conductivity can be read as

$$\sigma(\omega) = \frac{1}{(\nu - i\omega)}\frac{ne^2}{m} \tag{71}$$

In infrared range $\omega << \nu \sim 10^{14}(1/\sec)$, so the DC conductivity

$$\sigma = \frac{ne^2}{m\nu} \tag{72}$$

can be taken.

Let us now compare the magnitude of conduction current with that of the displacement current.

Assume $E = E_0 e^{-i\omega t}$. Then

$$\left|\frac{j_f}{\varepsilon\frac{\partial E}{\partial t}}\right| = \frac{\sigma E}{\varepsilon\omega E} = \frac{\sigma}{\varepsilon\omega} \tag{73}$$

In copper, $\sigma = 6 \times 10^7 (s/m)$. The condition for $j_f \approx \varepsilon\frac{\partial E}{\partial t}$, or $\frac{\sigma}{\varepsilon\omega} \approx 1$ leads to

$$\omega = \frac{\sigma}{\varepsilon} = \frac{6 \times 10^7}{8.85 \times 10^{-12}} \sim 7 \times 10^{19}(rad/\sec) \tag{74}$$

At frequencies $\omega < 10^{12}(rad/\sec)$ (communication wave frequency), $\frac{\sigma}{\varepsilon\omega} >> 1$ or $\left|j_f\right| >> \left|\varepsilon\frac{\partial E}{\partial t}\right|$.

12. Electromagnetic pulse and high power microwave overview

Several nations and terrorists have a capability to use electromagnetic pulse (EMP) as a weapon to disrupt the critical infrastructures. Electromagnetic pulse is an intense and direct energy field that can interrupt sensitive electrical and electronic equipment over a very wide area, depending on power of the nuclear device and altitude of the burst. An explosion exploded at few heights in the atmosphere can produce EMP and known as high altitude EMP or HEMP. High power microwave (HPM) can be produced with the help of powerful batteries by electrical equipment that transforms battery power into intense microwaves which may be harmful electronics equipments [62–71]. The high- power electromagnetic (HPEM) term describes a set of transient electromagnetic environments with intense electric and magnetic fields. High- power electromagnetic field may be produced by electrostatic discharge, radar system, lightning strikes, etc. The nuclear bursts can lead to the production of electromagnetic pulse which may be used against the enemy country's military satellites. Therefore the sources derived from lasers, nuclear events are vulnerable and called laser and microwave threats. Microwave weapons do not rely on exact knowledge of the enemy system. These weapons can leave persisting and lasting effects in the enemy targets through damage and destruction of electronic circuits, components. Actually HEMP or HPM energy fields, as they instantly spread outward, may also affect nearby hospital equipment or personal medical devices, such as pacemakers. These may damage critical electronic systems throughout other parts of the surrounding civilian infrastructure. HEMP or HPM may damage to petroleum, natural gas infrastructure, transportation systems, food production, communication systems and financial systems [62–71].

13. Generation of high - frequency instability through plasma environment

The beams of ions and electrons are a source of free energy which can be transferred to high power waves. If conditions are favorable, the resonant interaction of the waves in plasma can lead to nonlinear instabilities, in which all the waves grow faster than exponentially and attain enormously large amplitudes. These instabilities are referred to as explosive instabilities. Such instabilities could be of considerable practical interest, as these seem to offer a mechanism for rapid dissipation of coherent wave energy into thermal motion, and hence may be effective for plasma heating [72, 73]. A consistent theory of explosive instability shows that in the three-wave approximation the amplitudes of all the waves tend to infinity over a finite time called explosion time [74, 75]. In ref. [74], an explosive- generated – plasma is discovered for low and high frequency instabilities. The solution of dispersion equation is found numerically for the possibility of wave triplet and synchronism conditions. The instabilities is observed to propagate whose wave number.

14. Electron beam plasma model and theoretical calculation

Here we considers ions, electrons and negatively charged electron beam are immersed in a Hall thruster plasma channel [51–55]. The magnetic field is consider as $\vec{B} = B\hat{z}$ so that electrons are magnetized while ions remains un-magnetized and electrons rotates with cyclotron frequency $\Omega = \frac{eB}{m_e}$, whereas the gyro-radius for ions is larger so that they cannot rotate and simply ejects out by providing thrust to the device. The axial electric field $\vec{E} = E\hat{x}$ (along the x - axis) which accelerates the particles. It causes electrons have a $\vec{E} \times \vec{B}$ drift in the azimuthal direction (y-axis) whereas the movement of ions is restricted along x-axis. Similar to previous studies, here, we consider the motion of all the species i.e. for ions (density n_i, massm_i, velocity v_i) for electrons (density n_e, mass m_e, velocity v_e), for electron beam (density n_b, mass m_b, velocity v_b) and collision frequency for the excitation of instability. The basic fluid equations are given as follows:

$$\frac{\partial n_i}{\partial t} + \vec{\nabla} \cdot \left(\vec{v}_i n_i\right) = 0 \quad (75)$$

$$m_i\left(\frac{\partial}{\partial t}+\left(\vec{v}_i.\vec{\nabla}\right)\right)\vec{v}_i=e\vec{E} \tag{76}$$

$$\frac{\partial n_e}{\partial t}+\vec{\nabla}\cdot\left(\vec{v}_e n_e\right)=0 \tag{77}$$

$$m_e\left(\frac{\partial}{\partial t}+\left(\vec{v}_e\cdot\vec{\nabla}\right)+v\right)\vec{v}_e=-e\left(\vec{E}+\vec{v}_e\times\vec{B}\right) \tag{78}$$

$$\frac{\partial n_b}{\partial t}+\vec{\nabla}\cdot\left(\vec{v}_b n_b\right)=0 \tag{79}$$

$$m_b\left(\frac{\partial}{\partial t}+\left(\vec{v}_b\cdot\vec{\nabla}\right)\right)\vec{v}_b=-en_b\vec{E} \tag{80}$$

$$\varepsilon_0\nabla^2\varphi_1=e(n_{e1}-n_{i1}+n_{b1}) \tag{81}$$

Since the larmor radius of ions are larger than the length of the channel (6 cm), therefore ions are considered as unmagnetized in the channel and are accelerated along the axial direction of the chamber. We consider ions initial drift in the positive x – direction ($\vec{v}_{i0}=v_{i0}\hat{x}$) with neglecting motion in both azimuthal and radial directions [51–55]. Electron has motion in the x-direction ($\vec{v}_b=v_b\hat{x}$) since electrons are affected by magnetic field and get magnetized, we takes their $\vec{E}\times\vec{B}$ initial drift in the y – direction ($\vec{v}_e=v_e\hat{y}$).

To find the oscillations by the solutions of the above equations we take the quantities varied as the $A(r,t)=A_0e^{i(k.r-\omega t)}$ for first order perturb quantities $n_{i1},n_{e1},n_{b1},v_{i1},v_{e1},v_{b1}$ and $\vec{E}_1$ together with ω as a frequency of oscillations and the k is the wave propagation vector within plane of (x, y) . On remarking the magnetic fields are large enough in Hall thruster and condition $\Omega>>\omega,k_yv_{e0},v$ is satisfied [51–56]. By solving the equation of motion and the equation of continuity for electrons, we get the perturbed density of electrons in terms of oscillating potential φ_1 in the following way

$$n_{e1}=\frac{en_{e0}\hat{\omega}k^2\varphi_1}{m_e\Omega^2\left(\omega-k_yv_{e0}\right)} \tag{82}$$

Let us consider, $\hat{\omega}=\omega-k_yv_{e0}-iv$, the cyclotron frequency $\Omega=\frac{eB}{m_e}$ and $k^2=k_x^2+k_y^2$.

Similarly, on solving equation for ions we get the ion density term as

$$n_{i1}=\frac{ek^2n_{i0}\varphi_1}{m_i(\omega-k_xv_{i0})^2} \tag{83}$$

Similarly for electron beam density given as

$$n_{b1}=-\frac{ek^2n_{b0}\varphi_1}{m_b(\omega-k_xv_{b0})^2} \tag{84}$$

By putting these density values in the Poisson's equations

$$-k^2\varphi_1=\frac{e^2n_{e0}\hat{\omega}k^2\varphi_1}{m_e\varepsilon_0\Omega^2\left(\omega-k_yv_{e0}\right)}-\frac{e^2k^2n_{i0}\varphi_1}{m_i\varepsilon_0(\omega-k_xv_{i0})^2}-\frac{e^2k^2n_{b0}\varphi_1}{m_b\varepsilon_0(\omega-k_xv_{b0})^2} \tag{85}$$

On taking the plasma frequencies as; $\omega_{pe} = \sqrt{\frac{e^2 n_{eo}}{m_e \varepsilon_0}}$, $\omega_{pi} = \sqrt{\frac{e^2 n_{io}}{m_i \varepsilon_0}}$, and $\omega_{pb} = \sqrt{\frac{e^2 n_{bo}}{m_b \varepsilon_0}}$. Then the above equation reduces in the form as

$$-k^2 \varphi_1 = \frac{\omega_{pe}^2 \hat{\omega} k^2 \varphi_1}{\Omega^2 \left(\omega - k_y v_{e0}\right)} - \frac{\omega_{pi}^2 k^2 \varphi_1}{\left(\omega - k_x v_{i0}\right)^2} - \frac{\omega_{pb}^2 k^2 \varphi_1}{\left(\omega - k_x v_{b0}\right)^2} \tag{86}$$

Since the perturbed potential is not zero i.e. $\varphi_1 \neq 0$ then we get

$$\frac{\omega_{pe}^2 \hat{\omega}}{\Omega^2 \left(\omega - k_y v_{e0}\right)} - \frac{\omega_{pi}^2}{\left(\omega - k_x v_{i0}\right)^2} - \frac{\omega_{pb}^2}{\left(\omega - k_x v_{b0}\right)^2} + 1 = 0 \tag{87}$$

This is the modified dispersion relation for the lower-hybrid waves under the effects of collisions and electrons beam density.

15. Analytical solutions under the limitations

Consider now waves propagating along the $\hat{y}$ direction, so that $k_x = 0$, which, in real thruster geometry, corresponds to azimuthally propagating, waves. We discuss below its limiting cases through Litvak and Fisch [78].

$$\omega << \left|k_y v_{e0}\right|, \tag{88}$$

The solutions for the dispersion relation (57) can be obtained as follows:

$$\omega^2 \approx \frac{\left(\omega_{pi}^2 + \omega_{pb}^2\right)\Omega^2}{\left(\Omega^2 + \omega_{pe}^2\right)\left[1 + \frac{i\nu_e \omega_{pe}^2}{\left(\Omega^2 + \omega_{pe}^2\right) k_y v_{e0}}\right]} \tag{89}$$

Since the last terms in the second square brackets of the denominator in the right-hand side of (89) are small, we obtain the following

$$\omega \approx \pm \sqrt{\frac{\Omega^2 \left(\omega_{pi}^2 + \omega_{pb}^2\right)}{\left(\Omega^2 + \omega_{pe}^2\right)}} \left[1 - \frac{i\nu_e \omega_{pe}^2}{2k_y v_{e0}\left(\Omega^2 + \omega_{pe}^2\right)}\right] \tag{90}$$

Finally, the growth rate γ of the resistive instability is calculated from (90) as follow

$$\gamma \approx \frac{\nu_e \omega_{pe}^2}{2k_y v_{e0}\left(\Omega^2 + \omega_{pe}^2\right)} \times \sqrt{\frac{\Omega^2 \left(\omega_{pi}^2 + \omega_{pb}^2\right)}{\left(\Omega^2 + \omega_{pe}^2\right)}} \tag{91}$$

The corresponding real frequency $\omega_r (\omega \equiv \omega_r \pm i\gamma)$ is obtained as

$$\omega_r \approx \sqrt{\frac{\Omega^2 \left(\omega_{pi}^2 + \omega_{pb}^2\right)}{\left(\Omega^2 + \omega_{pe}^2\right)}} \tag{92}$$

The Eqs. (91) show that the growth of the high frequency instability depends on collision frequency, electron density, ion density, beam density, azimuthal wave

Parameters	Range
Magnetic field	$B_{0z} \sim 100 - 200G$
Axial Wave number	Kx $\sim$ 200- 600/m
Azimuthal Wave number	Ky $\sim$ 400-1200/m
Collisional frequency	$v{\sim}10^6/s$
Initial drift of electron	$u_0{\sim}10^6 m/s$
Initial drift of ions	$v_0 \sim 2 \times 10^4 - 5 \times 10^4 m/s$
Plasma density	$n_{e0} \sim 10^{18}, n_{i0} \sim 10^{18}, n_{b0} \sim 10^{17}/m^3$
Thruster channel diameter	D $\sim$ 4 – 10 cm

Table 2.
Plasma parameters.

number, initial drift and on the applied magnetic field. On the other hand, the real frequency of the wave depends only on the magnetic field, electron plasma density, ion density and beam density. By tuning these parameters one can control the frequency of the generating wave. In the below **Table 2**, the different parameters of a Hall thruster are given [51–56].

16. Results and discussion

The Eqs. (91) and (92) are solved with MATLAB by using appropriate parameters given in **Table 2**. We plot various figures for investigating the variation of growth rate and real frequency of the instability with magnetic field B_0 and density of beam n_b, initial drift, collision frequency ν and wave number. For these sets of parameters, only one dominated mode of the dispersion relation is plotted in the figures. **Figure 3** shows the variation of growth rate and real frequency for different values of magnetic field. The reason for the enhanced growth rate as well as real frequency can be understood based on Lorentz force and the electron collisions. Since the electrons have their drift in the y-direction, they experience the Lorentz force due to the magnetic field in the negative of x-direction, i.e., in the direction opposite to the ions drift. The higher Lorentz force helps these transverse oscillations to grow relatively at a faster rate owing to an enhancement in the frequency. On the other hand, this is quite plausible that larger cyclotron frequency of the electrons leads to stronger effects of the collisions because of which the resistive coupling becomes more significant and hence the wave grows at its higher rate. Opposite effect of the magnetic field was observed by Alcock and Keen in case of a drift dissipative instability that occurred in afterglow plasma [76]. Similarly studied are also investigated by Sing and Malik in magnetized plasma [51–56].

In **Figure 4**, we have plotted the variation of growth rate γ and real frequency with the azimuthal wavenumber in order to examine the growth of these waves, when the oscillations are of smaller or relatively longer wavelengths. Here, the oscillations of larger wave numbers (or smaller wavelengths) are found to have lower growth. The faster decay that is observed on the larger side of k is probably due to the stronger Landau damping. The growth rate shows parabolic nature but the real frequency is almost increases linearly with respect to azimuthal wave number. It means that oscillations of smaller wavelengths are most unstable. Kapulkin et al. have theoretically observed the growth rate of instability to directly

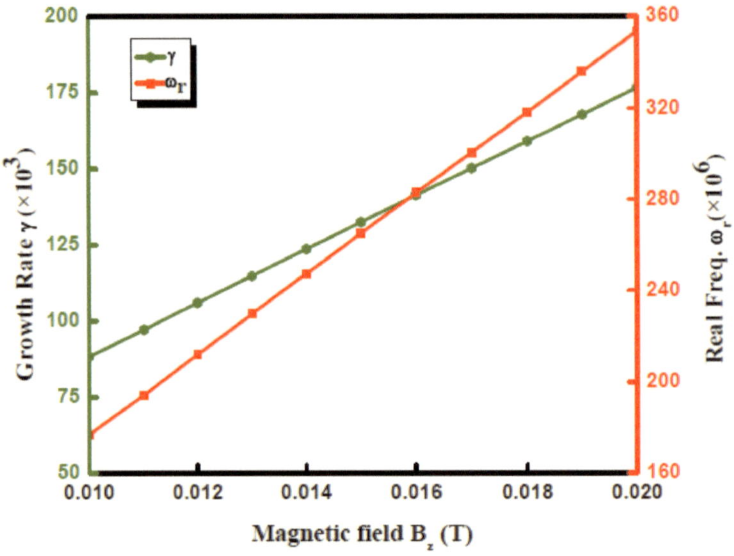

Figure 3.
Variation of growth rate and real frequency with the magnetic field.

proportional to the azimuthal wavenumber [77]. Litvak and Fisch have also shown that the rate of growth of instability is inversely proportional to the azimuthal wave number [78].

On the other hand, the variation of growth rate γ and real frequency with the collision frequency is depicted in **Figure 5**. The wave grow at faster rates in the presence of more electron collisions. This is due to the resistive coupling, which get much stronger in the presence of more collisions. In the present case, the growth rate grows at a much faster rate and real frequency is constant, and graph shows that the growth rate is directly proportional to the collision frequency. During the simulation studies of resistive instability, Fernandez *et al.* also observed the growth

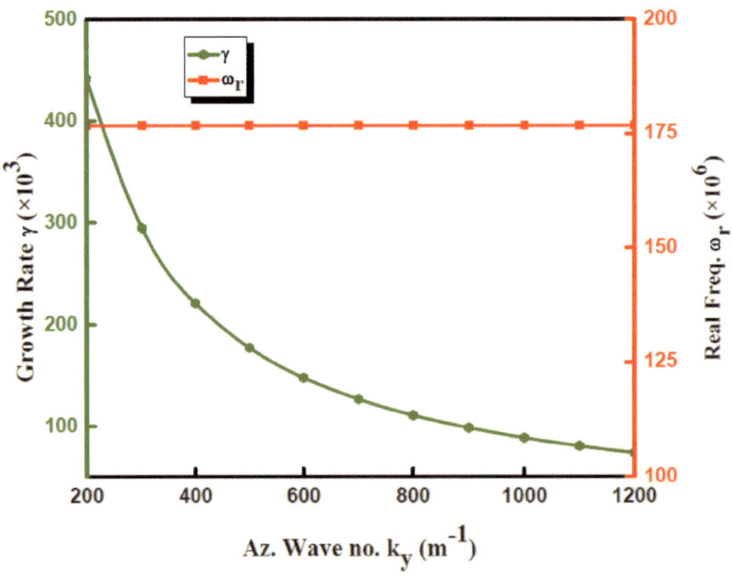

Figure 4.
Variation of growth rate γ and real frequency with azimuthal wavenumber.

rate to be directly proportional to the square root of the collision frequency [79]. In **Figure 6,** we show the dependence of the growth rate on the electron drift velocity. it is observed that the growth rate is reduced in the presence of larger electron drift velocity. In this case the resistive coupling of the oscillations to the electrons' drift would be weaker due to the enhanced velocity of the electrons. The reduced growth under the effect of stronger magnetic field is attributed to the weaker coupling of the oscillations to the electrons closed drift. The variation of growth rate γ and real frequency with beam density are shown in **Figure 7**. The growth shows asymmetric Gaussian type behavior but the real frequency varies linearly with beam density of electrons. This is due to the increased collisional effect with the large plasma density.

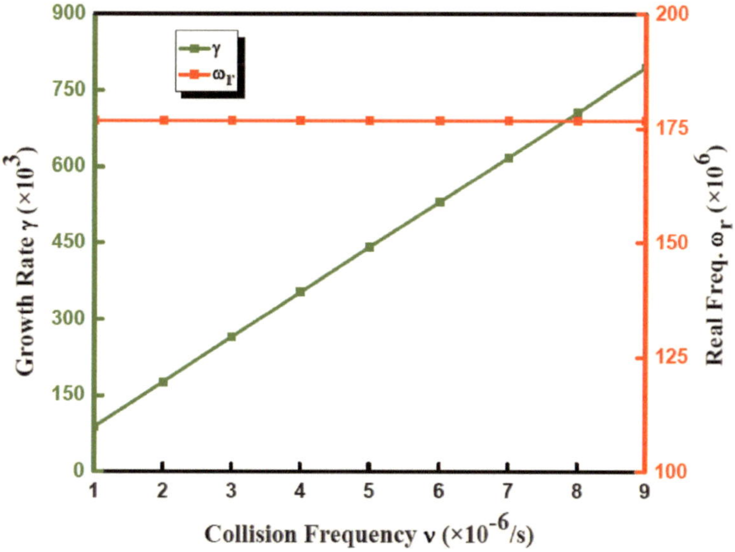

Figure 5.
Variation of growth rate γ and real frequency with collision frequency.

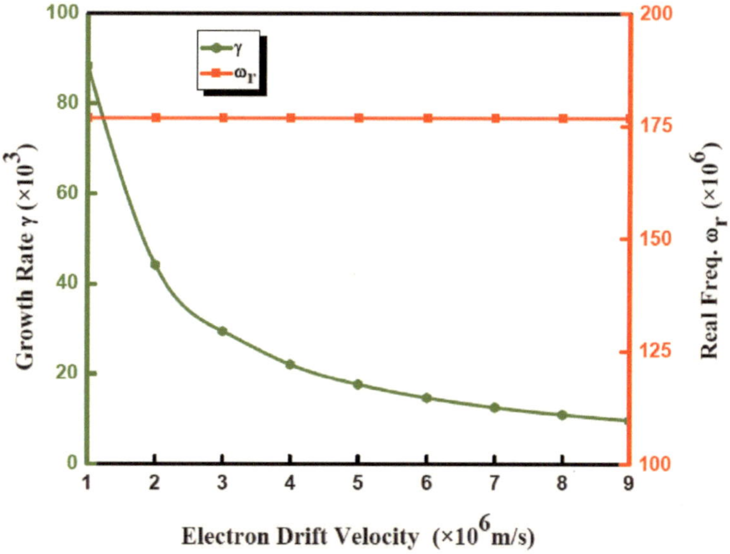

Figure 6.
Variation of growth rate γ and real frequency with electron drift velocity.

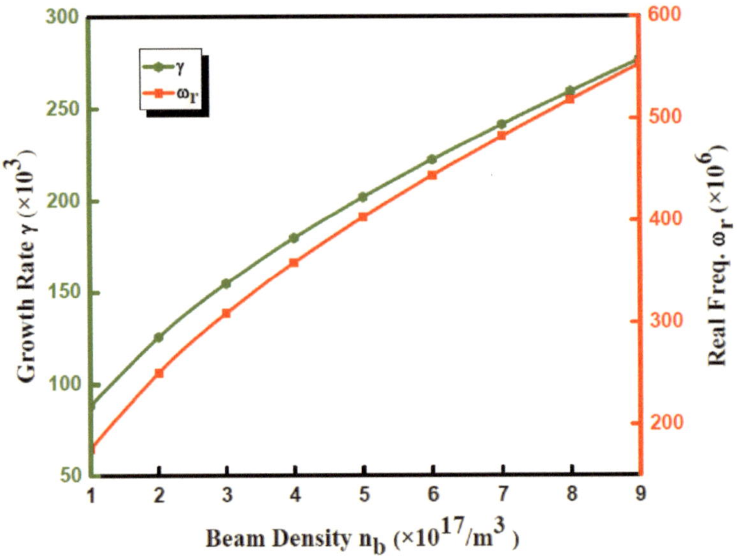

Figure 7.
Variation of growth rate γ and real frequency with beam density.

17. Conclusions

The present chapter discuss the properties of electromagnetic waves propagating through different media. In the first part of the chapter, the dispersion relation for the electromagnetic waves in conducting medium is derived. It has been experienced that the penetration of the electromagnetic field depend on the frequency of the source as well as the electrical properties of the medium. The significance of skin depth for biological and conducting media are explained through numerical examples. In the second part of the chapter, the generation of high frequency instability in plasma is discussed which grow with the magnetic field, wave length, collision frequency and the beam density. The growth rate linearly increases with collision frequency of electrons but it is decreases with the drift velocity of electrons. The real frequency of the instability increases with magnetic field, azimuthal wave number and beam density. The real frequency is almost independent with the collision frequency of the electrons.

Acknowledgements

The University Grants Commission (UGC), New Delhi, India is thankfully acknowledged for providing the startup Grant (No. F. 30-356/2017/BSR).

Author details

Sukhmander Singh[1*], Ashish Tyagi[2] and Bhavna Vidhani[3]

1 Plasma Waves and Electric Propulsion Laboratory, Department of Physics, Central University of Rajasthan, Ajmer, Kishangarh, India

2 Physics Department, Swami Shraddhanand College, University of Delhi, Delhi, India

3 Department of Physics and Electronics, Hansraj College, University of Delhi, Delhi, India

*Address all correspondence to: sukhmandersingh@curaj.ac.in

References

[1] L. Liu, S. Zhou and J. Cui, “Prospects and Problems of Wireless Communication for Underwater Sensor Net-works,” Wiley WCMC Special Issue on Underwater Sen-sor Networks (Invited), 2008.

[2] J. H. Goh, A. Shaw, A. I. Al-Shanmma’a, “Underwater Wireless Communication System,” Journal of Physics, Conference Series 178, 2009.

[3] A. I. Al-Shamma’a, A. Shaw and *S. saman*, “Propagation of Electromagnetic Waves at MHz Frequencies through Seawater,” IEEE Transactions on Antennas and Propa- tion, Vol. 52, No. 11, November 2004, pp. 2843-2849.

[4] S. Bogie, “Conduction and Magnetic Signaling in the Sea,” Radio Electronic Engineering, Vol. 42, No. 10, 1972, pp. 447-452. doi:10.1049/ree.1972.0076

[5] Shan Jiang, Stavros Georgakopoulos, Electromagnetic Wave Propagation into Fresh Water, Journal of Electromagnetic Analysis and Applications, 2011, 3, 261-266

[6] Habib Mazharimousavi S, Roozbeh A, Halilsoy M. Electromagnetic wave propagation through inhomogeneous material layers. Journal of Electromagnetic Waves and Applications. 2013 Nov 1;27(16): 2065-74.

[7] Andrey B. Matsko and Yuri V. Rostovtsev, Electromagnetic-wave propagation and amplification in overdense plasmas: Application to free electron lasers, Physical Review E, 58, 6, DECEMBER 1998

[8] P. Sprangle, E. Esarey, J. Krall, and G. Joyce, Phys. Rev. Lett. 69, 2200 ~1992.

[9] .C. Oldano, Electromagnetic-wave propagation in anisotropic stratified media, Physical Review A, 40, 10 NOVEMBER 15, 1989

[10] Hidenori Akiyama, Takashi Sakugawa, Takao Namihira, Industrial Applications of Pulsed Power Technology, IEEE Transactions on Dielectrics and Electrical Insulation Vol. 14, No. 5; October 2007, 1051-1064

[11] A. Pokryvailo, M. Wolf, Y. Yankelevich, S. Wald, L. R. Grabowski, E. M.van Veldhuizen, Wijnand R. Rutgers, M. Reiser, B. Glocker, T. Eckhardt, P. Kempenaers and A. Welleman, “High-Power Pulsed Corona for Treatmentof Pollutants in Heterogeneous Media”, IEEE Trans. Plasma Sci., Vol. 34,pp. 1731-1743, 2006.

[12] G.J.J. Winands, K. Yan, A.J.M. Pemen, S.A. Nair, Z. Liu and E.J.M. vanHeesch, “An Industrial Streamer Corona Plasma System for Gas Cleaning”,IEEE Trans. Plasma Sci., Vol. 34, pp.2426-2433, 2006.

[13] W. Hartmann, T. Hammer, T. Kishimoto, M. Romheld and A. Safitri, “Ozone Generation in a Wire-Plate Pulsed Corona Plasma Reactor” , 15th IEEE Pulsed Power Conf., pp. 856-859, 2005.

[14] J. Choi, T. Yamaguchi, K. Yamamoto, T. Namihira, T. Sakugawa, S. Katsuki and H. Akiyama, “Feasibility Studies of EMTP Simulation for the Design of the Pulsed-Power Generator Using MPC and BPFN for Water Treatments”, IEEE Trans. Plasma Sci., Vol. 34, pp.1744-1750, 2006.

[15] R.W. Bruce, R.L. Bruce, A.W. Fliflet, M. Kahn, S.H. Gold, A.K. Kinkead, D. Lewis, III and M.A. Imam, "Joining of ceramic tubes using a high-power 83-GHz Millimeter-wave beam", IEEE Trans. Plasma Sci., Vol. 33, pp. 668-678, 2005.

[16] A. M. Loske, "Medical and Biomedical Applications of Shock Waves: The State of the Art and the Near Future," in 30th International Symposium on Shock Waves 1, 2017, pp. 29-34.

[17] Claire M. Watts, Xianliang Liu, and Willie J. Padilla, Metamaterial Electromagnetic Wave Absorbers, Adv. Mater. 2012, 24, OP98–OP120, DOI: 10.1002/adma.201200674

[18] A. I. M. Ayala , Master of Science Thesis, Tufts University, USA, 2009 .

[19] Y. Gong , Z. Li , J. Fu , Y. Chen , G. Wang , H. Lu , L. Wang , X. Liu , Opt. Exp. 2011 , 19 , 10193

[20] Z. H. Jiang , S. Yun , F. Toor , D. H. Werner , T. S. Mayer , ACS Nano 2011 , 5 , 4641 .

[21] X. J. He , Y. Wang , J. M. Wang , T. L. Gui , PIER 2011 , 115 , 381

[22] E. Berry et al., Optical properties of tissue measured using terahertzpulsed imaging," Proc. SPIE, vol. 5030, pp. 459-470, Jun. 2003.

[23] A. J. Fitzgerald et al., Catalogue of human tissue optical properties at terahertz frequencies, J. Biol. Phys., vol. 29, nos. 2-3, pp. 123-128, 2003.

[24] G. W. Hanson, Dyadic Green's functions and guided surface waves for a surface conductivity model of graphene, J. Appl. Phys., vol. 103, no. 6, pp. 064302-1–064302-8, Mar. 2008.

[25] J. M. Jornet and I. F. Akyildiz, Graphene-based nano-antennas for electromagnetic nanocommunications in the terahertz band, in Proc. 4th Eur. Conf. Antennas Propag. (EUCAP), Apr. 2010, pp. 1-5.

[26] T. Binzoni, A.Vogel, A. H. Gandjbakhche, and R. Marchesini, Detection limits of multi-spectral optical imaging under the skin surface, Phys. Med. Biol., vol. 53, no. 3, pp. 617-636, 2008.

[27] Loske, A.M.: Medical and Biomedical Applications of Shock Waves. Shock Wave and High Pressure Phenomena. Springer International Publishing AG, Cham, Switzerland (2017). ISBN 978-3-319-47568-4

[28] Wang, C.J.: Extracorporeal shockwave therapy in musculoskeletal disorders. J. Orthop. Surg. Res. 7, 11–17 (2012)

[29] Kenmoku, T., Nobuyasu, O., Ohtori, S.: Degeneration and recovery of the neuromuscular junction after application of extracorporeal shock wave therapy. J. Orthop. Res. 30, 1660–1665 (2012)

[30] Delius, M., Hofschneider, P.H., Lauer, U., Messmer, K. Extracorporeal shock waves for gene therapy? Lancet 345, 1377 (1995)

[31] Chen ZY, Gao YM, Du M. Propagation characteristics of electromagnetic wave on multiple tissue interfaces in wireless deep implant communication. IET Microwaves, Antennas & Propagation. 2018 Jul 6;12 (13):2034-40.

[32] Callejon, M.A., Naranjo-Hernandez, D., Reina-Tosina, J. and Roa LM, Distributed circuit modeling of galvanic and capacitive coupling for intrabody communication', IEEE Trans. Biomed. Eng., 2012, 59, (11), pp. 3263– 3269

[33] Seyedi M, Kibret B, Lai DT and Faulkner M. A survey on intrabody communications for body area network applications', IEEE Trans. Biomed. Eng., 2013, 60, (8), pp. 2067– 2079

[34] Marani R, Perri AG. RFID technology for biomedical applications: State of art and future developments. i-Manager's Journal on Electronics Engineering. 2015 Dec 1;6(2):1.

[35] A. Denisov and E. Yeatman, "Ultrasonic vs. inductive power delivery for miniature biomedical implants," in Body Sensor Networks (BSN), 2010 International Conference on, 2010, pp. 84-89: IEEE.

[36] S. Ozeri and D. Shmilovitz, "Ultrasonic transcutaneous energy transfer for powering implanted devices," Ultrasonics, vol. 50, no. 6, pp. 556-566, 2010.

[37] Li C, Un KF, Mak PI, Chen Y, Muñoz-Ferreras JM, Yang Z, Gómez-García R. Overview of recent development on wireless sensing circuits and systems for healthcare and biomedical applications. IEEE Journal on Emerging and Selected Topics in Circuits and Systems. 2018 Apr 3;8(2): 165-77.

[38] R. Muller et al., "A minimally invasive 64-channel wireless μECoG implant," IEEE Journal of Solid-State Circuits, vol. 50, no. 1, pp. 344-359, 2015.

[39] G. Papotto, F. Carrara, A. Finocchiaro, and G. Palmisano, "A 90nm CMOS 5Mb/s crystal-less RF transceiver for RF-powered WSN nodes," in Solid-State Circuits Conference Digest of Technical Papers (ISSCC), 2012 IEEE International, 2012, pp. 452-454: IEEE.

[40] Z.-P. Liang and P. C. Lauterbur, Principles of magnetic resonance imaging: a signal processing perspective. SPIE Optical Engineering Press, 2000.

[41] P. Lauterbur, "Image formation by induced local interactions: examples employing nuclear magnetic resonance," 1973.

[42] S. Gabriel, R. Lau, and C. Gabriel, "The dielectric properties of biological tissues: III. Parametric models for the dielectric spectrum of tissues," Physics in medicine and biology, vol. 41, no. 11, p. 2271, 1996.

[43] C. Gabriel, S. Gabriel, and E. Corthout, "The dielectric properties of biological tissues: I. Literature survey," Physics in medicine and biology, vol. 41, no. 11, p. 2231, 1996.

[44] S. Gabriel, R. Lau, and C. Gabriel, "The dielectric properties of biological tissues: II. Measurements in the frequency range 10 Hz to 20 GHz," Physics in medicine and biology, vol. 41, no. 11, p. 2251, 1996.

[45] Johnson C C and Guy AW, Nonionizing Electromagnetic Wave Effects in Biological Materials and Systems, Proceedings of THE IEEB, VOL. 60, NO. 6, J ~ T E(19 72)692-718

[46] C. J. Restall, P. F. Leonard, H. F. Taswell, and R. E. Holaday. IMPI Symp. (Univ. of Alberta, Edmonton, Canada, May 21-"Warming of human blood by use of microwaves," in Summ. 4th, 23, 1969), pp. 9699.

[47] Inan US, Said RK and Inan AS. Engineering electromagnetics and waves. Pearson; 2014 Mar 14.

[48] David J. Griffiths Introduction to Electrodynamics. (Addison-Wesley: Upper Saddle River, 1999).

[49] Jackson, J. D. (1998). Classical Electrodynamics. New York: Wiley, 3rd edition

[50] Mallick C, Bandyopadhyay M, Kumar R. Evolution of Microwave Electric Field on Power Coupling to Plasma during Ignition Phase. InSelected Topics in Plasma Physics 2020 Jun 15. IntechOpen.

[51] Tyagi J, Singh S, Malik HK, Effect of dust on tilted electrostatic resistive instability in a Hall thruster. Journal of Theoretical and Applied Physics. 2018; 12: 39-43. Doi.org/10.1007/s40094-018-0278-z

[52] Singh S, Malik H K, Nishida Y. High frequency electromagnetic resistive

instability in a Hall thruster under the effect of ionization. Physics of Plasmas.2013; 20: 102109 (1-7).

[53] Singh S, Malik H K. Growth of low frequency electrostatic and electromagnetic instabilities in a Hall thruster. IEEE Transactions on Plasma Science.2011; 39:1910-1918.

[54] Singh S, Malik H K. Resistive instabilities in a Hall thruster under the presence of collisions and thermal motion of electrons. The Open Plasma Physics Journal. 2011; 4:16-23.

[55] Malik H K and Singh S. Resistive instability in a Hall plasma discharge under ionization effect. Physics of Plasmas.2013; 20: 052115 (1-8).

[56] Singh S. Evolutions of Growing Waves in Complex Plasma Medium. In edited book Engineering Fluid Mechanics. IntechOpen, London, United Kingdom, Nov 2020

[57] Singh S. Waves and Instabilities in E X B Dusty Plasma. In the edited book Thermophysical Properties of Complex Materials. IntechOpen, London, United Kingdom, December 12th 2019

[58] Singh S. Dynamics of Rayleigh-Taylor Instability in Plasma Fluids. In the edited book Engineering Fluid Mechanics. IntechOpen, London, United Kingdom, April 15th 2020

[59] Singh S. Hall Thruster: An Electric Propulsion through Plasmas. In the edited book Plasma Science IntechOpen, London, United Kingdom, March 2nd 2020 Doi.org/10.1063/1.2823033

[60] Singh S, Kumar S, Sanjeev, Meena S K and Saini S K. Introduction to Plasma Based Propulsion System: Hall Thrusters. In the edited book Propulsion - New Perspectives and Applications" edited by Prof. Kazuo Matsuuchi, IntechOpen, London, United Kingdom, March 2021

[61] Singh S, editor. Selected Topics in Plasma Physics. BoD–Books on Demand; 2020 Nov 19.

[62] Khalatpour A, Paulsen AK, Deimert C, Wasilewski ZR, Hu Q. High-power portable terahertz laser systems. Nature Photonics. 2021 Jan;15(1):16-20.

[63] Wu S, Cui S. Overview of High-Power Pulsed Power Supply. InPulsed Alternators Technologies and Application 2021 (pp. 1-35). Springer, Singapore.

[64] Wu S, Cui S. Electromagnetic Weapon Load of Pulsed Power Supply. InPulsed Alternators Technologies and Application 2021 (pp. 209-227). Springer, Singapore.

[65] Radasky, W. A., C. E., Baum, Wik, M. W.: Introduction to the special issue on high power electromagnetics (HPEM) and intentional electromagnetic interference (IEMI) environments and test capabilities. IEEE Trans. Electromagn. Compat. 46 (2004)

[66] Sabath, F., Backstrom, M., Nordstrom, B., Serafin, D., Kaiser, A., Kerr, B., Nitsch, D.: Overview of four European high power microwave narrow band test facilities. IEEE Trans. Electromagn. Compat. 46, 329 (2004)

[67] Parfenov, Y. V., Zdoukhov, L. N., Radasky, W. A., Ianoz, M.: Conducted IEMI threats for commercial buildings. IEEE Trans. Electromagn. Compat. 46, 404 (2004)

[68] Lu X, Picard JF, Shapiro MA, Mastovsky I, Temkin RJ, Conde M, Power JG, Shao J, Wisniewski EE, Peng M, Ha G. Coherent high-power RF wakefield generation by electron bunch trains in a metamaterial structure. Applied Physics Letters. 2020 Jun 29;116 (26):264102.

[69] Zhang J, Zhang D, Fan Y, He J, Ge X, Zhang X, Ju J, Xun T. Progress in

narrowband high-power microwave sources. Physics of Plasmas. 2020 Jan 17; 27(1):010501.

[70] Frank JW. Electromagnetic fields, 5G and health: what about the precautionary principle?. J Epidemiol Community Health. 2021 Jun 1;75(6): 562-6.

[71] Wu S, Cui S. Basic Theories of Pulsed Alternators. InPulsed Alternators Technologies and Application 2021 (pp. 37-61). Springer, Singapore.

[72] Aamodt, R. E., Sloan, M. L.: Nonlinear interactions of positive and negative energy waves. Phys. Fluids 11, 2218 (1968)

[73] Wilhelmson, H., Stenflo, I., Engelmann, F.: Explosive instabilities in the well defined phase description. J. Math. Phys. 11, 1738 (1970)

[74] O.P. Malik, Sukhmander Singh, Hitendra K. Malik, A. Kumar. Low and high frequency instabilities in an explosion- generated-plasma and possibility of wave triplet. Journal of Theoretical and Applied Physics (2015) Vol. 9 Pgs.75 -80.

[75] O.P. Malik, Sukhmander Singh, Hitendra K. Malik, A. Kumar. High frequency instabilities in an explosion-generated-relativistic-plasma. Journal of Theoretical and Applied Physics (2015) Vol. 9, Pgs.105-110.

[76] Alcock M W & Keen B E, Phys Rev A 3, (1971) 1087.

[77] Kapulkin A, Kogan A & Guelman M, Acta Astronaut, 55 (2004) 109.

[78] Litvak A A & Fisch N J, Phys Plasmas, 8 (2001) 648.

[79] Fernandez E, Scharfe MK, Thomas CA, Gascon N & Cappelli MA, Phys Plasmas, 15, 012102 (2008).

Section 3

Electromagnetic Wave Propagation, Radiation and Scattering Applications

Chapter 6

Electromagnetism of Microwave Heating

Rafael Zamorano Ulloa

Abstract

Detailed electrodynamic descriptions of the fundamental workings of microwave heating devices are presented. We stress that all results come from Maxwell equations and the boundary conditions (BC). We analyze one by one the principal components of a microwave heater; the cooking chamber, the waveguide, and the microwave sources, either klystron or magnetron. The boundary conditions at the walls of the resonant cavity and at the interface air/surface of the food are given and show how relevant the BC are to understand how the microwaves penetrate the nonconducting, electric polarizable specimen. We mention the application of microwaving waste plastics to obtain a good H_2 quantity that could be used as a clean energy source for other machines. We obtained trapped stationary microwaves in the resonant cavity and traveling waves in the waveguides. We show 3D plots of the mathematical solutions and agree quite well with experimental measurements of hot/cold patterns. Simulations for cylindrical cavities are shown. The radiation processes in klystrons and magnetrons are described with some detail in terms of the accelerated electrons and their trajectories. These fields are sent to the waveguides and feed the cooking chamber. Whence, we understand how a meal or waste plastic, or an industrial sample is microwave heated.

Keywords: microwave heating, resonant cavity, cooking chamber, waveguide, klystron, magnetron, boundary conditions, food-air interface, Lienard-Wiechert potentials, Jefimenko fields

1. Introduction

Microwaves are everywhere and permeate the universe. They reach earth constantly and we produce them in many medical, industrial, chemical, domestic, and research on magnetic and dielectric materials and in devices and equipment [1–13]. The modern communications technology uses them intensely, Wi-Fi, all around the world, every second, every day [14]. Many medical applications are concentrated in cancer treatments by giving hyperthermia to the cancerous cells while avoiding to damage healthy cells. Microwave ablation is widely used in many types of cancers, bone, cardiac arrhythmic tissue, thyroid glands, skin cancer, and many other damaged tissues [2, 3, 15, 16]. The apparatus is constituted basically by a microwave source, a waveguide that ends in an antenna that, as a needle, penetrates the tissue [2, 3, 15, 16]. Industrial applications go from thermally treating/curing polymers, rubber, and plastics to quickly heat cement and minerals, and to assist vulcanization [4, 5, 17, 18]. Chemical applications are mainly directed to organic and/or inorganic synthesis and accelerating reactions, and to search for novel synthesis routes and novel products. Chemical microwave heating has been used for decades [6, 7]. Research in magnetics and in dielectrics includes heat transfer and/or electromagnetic excitation of matter [10–13]. Domestic technology uses them to heat up quickly and easily food, coffee, and water in microwave ovens (MWO) built for such function, see **Figure 1(D)** [8, 9]. By far, the two most commonly used microwave configurations include a source of microwaves, typically a klystron or a magnetron, a waveguide for these microwaves, and a resonant chamber where the microwaves are used to treat, to modify, to cure, to excite, or to heat up a sample put in the microwave chamber. There are two most common geometries of microwave chambers, cylindrical and rectangular [4–13]. **Figure 1(A)** shows an electric field pattern simulated inside a typical cylindrical cavity (3.4 cm of radius and 4.22 cm of height) used in a research equipment, in which magnetic samples are resonantly excited [19]. The magnetic field (not shown) is vertical (and orthogonal to E) and mostly concentrated along the z-axis at $r = 0$ and to its close vicinity. **Figure 1(B)** shows an electron paramagnetic resonance spectrometer that uses a cylindrical microwave chamber feed by a rectangular waveguide that collects the 9.4-GHz low-power microwaves produced by a klystron inside the box-labeled microwave bridge. **Figure 1(C)** is a calculated stationary electric field pattern from the solutions to Maxwell equations found in this work for a rectangular cavity. **Figure 1(D)** shows a typical domestic microwave oven of dimensions 26 cm × 30 cm × 34 cm.

A universal advantage of microwave heating in industrial, medical, chemical, and domestic processes is that it does it quickly and efficiently. Yet, several investigations are pursued to find even faster and better microwave heating schemes and profiles [15–17].

The three main parts of these heating devices constitute a source, a waveguide, and a heating chamber. **Figure 2(A)** shows the essential parts of a microwave oven commonly used to heat food. In addition to the already large number of industrial microwave heating applications, very recently, microwaving plastic waste decomposition has been proposed as a central step in order to generate clean hydrogen, H_2, out of heating a one-to-one mixture of triturated waste plastics with the catalyst FeAlOx [20]. Edwards et al. [21] have used microwaves to transform waste plastic bags, milk empty bottles, and other supermarket waste plastics, **Figure 2(B)**, in a clean hydrogen energy source. A 1:1 mixture of the catalyst FeAlOx and waste plastics heated up with microwaves in a cylindrical cavity

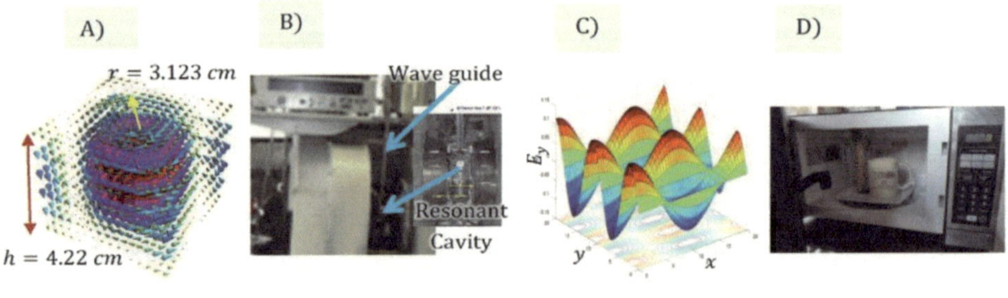

Figure 1.
Microwaves in cylindrical and rectangular geometries. (A) Simulation of the electric field, circular, pattern formed inside a cylindrical microwave cavity. (B) The microwave bridge, the rectangular waveguide, and a cylindrical cavity of a commercial electron paramagnetic resonance spectrometer used to excite magnetic samples. Here, heating is not desired, and the microwave power used is 1 mW or less. (C) A stationary electric field pattern calculated in this work from the solutions to Maxwell equations. The pattern is calculated for the planes x *and* y *with the coordinate* z *maintained fixed at an arbitrary height. (D) A typical domestic microwave oven (MWO), open and showing its internal chamber height* h = *38 cm, and width* a = *32 cm and* b = *depth, 30 cm.*

Figure 2.
Microwave heating. (A) A typical 1000 Watts, 2.45-GHz domestic microwave oven (MWO) cooking a chicken. (B) Plastic pollution derived from tons and tons of waste plastic bottles and supermarket bags. (C) Clean production of H_2 by microwaving waste plastic bags and bottles mixed 1:1 with the catalyst FeAlOx within a cylindrical cavity, the input power is 1000 Watts, and H_2 gas is liberated and carried toward a separate container. The cylindrical cavity operates in the TM_{010} mode.

(as shown in **Figure 2(C)**) for 10, 30, and 60 s extracts 85–90% of molecular hydrogen that is sent through a column to be stored in a separate chamber for its eventual utilization as a clean energy source. The principle of operation is the same for the domestic microwave oven with a rectangular heating chamber and for this microwave heater that is used to transform waste plastics within a cylindrical heating chamber. This transformation process is clean and fast and could help to reduce drastically the world's wide plastic contamination problem. Plastics invade mountains [22], forests, lakes, oceans, and cities [20–22]. As the hydrogen density content in plastic bags is around 14% per weight, its transformation into H_2 and multifaceted fullerenes might offer an opportunity of clean energy production for the countries interested in producing clean H_2 as a viable energy source for industrial and domestic usage and contribute to slowing down the global climate change [23]. Using tons and tons of garbage plastics, H_2 can be produced in high quantities and then used as a clean industry and house energy supply, this way contributing to combat the global warming. We consider this kind of potential application relevant for the plastic industry and pollution problems and much more relevant for its contribution to a cleaner, greener planet. Microwave heating is used globally, also, to cook or just heat up meals, water, coffee, and pizza. These microwave ovens operate at a power of 1000 Watts; the meals are put into a rectangular resonant cavity that is the cooking chamber that gives them their familiar "box" appearance. In this box, microwaves are delivered unevenly (see uneven electric field pattern calculated here and shown in **Figure 1(C)**), to the meals, and they readily penetrate the matter, making electric dipoles (mainly from water and fatty molecules) to oscillate frenetically; then, this excitation energy is passed to the rest of the specimen as heat.

In a few seconds or minutes, the specimen is hotter than at the start of the microwaving [5–13]. The utility of microwaves is multiple and of large range. In spite of the generalized use of microwave ovens, an informal survey has indicated to us that more than 90% of our STEM students do not really know how MWOs operate. The basic physics required to understand their workings is completed by the end of the undergraduate class work. This is telling us the degree of sophistication and depth that this technology carries, just as many other modern technologies do, as the Wi-Fi itself. A primal objective of this chapter is to describe its electromagnetic physics and to give the fundamentals. All comes from Maxwell equations. We emphasize the fact that most of the microwave heating devices (as the two devices shown in **Figure 2**) are composed of three essential electromagnetic parts: the production of the microwaves, their wave-guided structure that brings the microwaves from their origins toward the resonant cavity, being the third essential component, that is, the cooking chamber. Aforesaid, some devices use rectangular resonant cavities and others use cylindrical resonant cavities. We focus on rectangular cavities and just mention some results for cylindrical cavities.

We want to emphasize that the study of the electromagnetic functioning of a domestic, an industrial, or a laboratory MWO is an ideal technological case to see how the whole of the theory is applied. These microwave "boxes" contain all of the fundamental physics required to produce (in klystrons or magnetrons) microwaves, then input them into a loss-less waveguide, and finally get them to bath a rectangular heating chamber, or a cylindrical chamber, without appreciable microwave radiation being absorbed at the metallic walls. Their absorption is mainly carried out, precisely, by the specimen we want to heat up.

Our discussion has five parts: First, in Section 2, we deal qualitatively, in detail, with the fundamental constituents of a microwave heating system and the physical processes involved. Then, in Section 3, we analyze the boundary conditions, and in Section 4, we treat mathematically the resonant cavity. A description of the physics of this resonant cavity that keeps confined the microwaves all the time necessary for the food to warm up, or even be cooked, is given. We show that standing wave patterns are the solutions to Maxwell equations. Section 5 treats the rectangular waveguide and the general form of the traveling waves in them is obtained. The process of wave guiding the microwaves is the one that carries the microwaves from its source to the heating chamber. Then, in Section 7, we treat the production of microwaves in klystrons and/or magnetrons by radiating, accelerated, electrons moving in straight lines, or curved trajectories.

2. Fundamental constituents of a microwave heater

In formal terms, a domestic microwave oven [8, 9], or an industrial system [4, 5] or a laboratory prototype for clean extraction of H_2 from microwaving waste plastics [20, 21], is constituted by three fundamental parts shown in **Figure 2(A)** and **(C)** and in **Figure 3**.

(A) The resonant cavity: Once the microwaves are inside the rectangular cooking chamber, **Figure 2(A)**, or inside a cylindrical resonant cavity, **Figure 2(C)**, these microwaves display themselves stationary wave patterns since they are confined within good conducting walls, see **Figure 1(A)** and **(C)**. These waves rebound incessantly from these metallic walls without, practically, any energy loss. These wave patterns are specific for each geometry and each set of dimensions and the boundary conditions at the walls, as we show below.

In MWOs, the sixth wall is the see-through door that allows access to the interior of the chamber. The see-through window is covered by a metallic mesh with many

A) B)

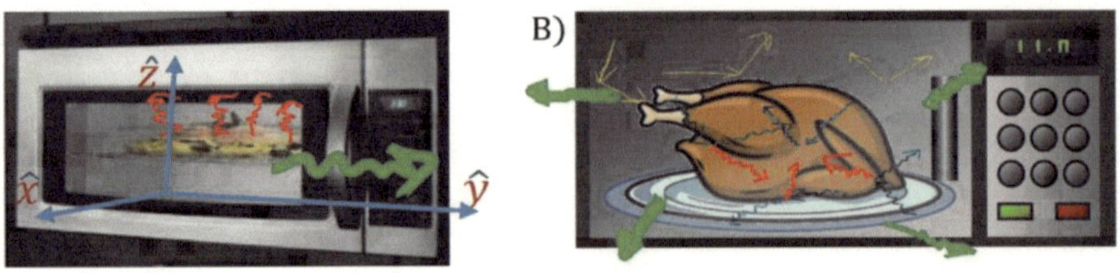

Figure 3.
Food being cooked in a microwave oven. (A) The meal is already hot and steam is actually getting out of the meat, and the green wavy arrow is pointing to the possibility that some microwaves get out of the oven. (B) The microwaves rebound from all six metallic surfaces of the "box" (yellow) and are reflected and transmitted from the surface of the specimen been heated (blue and red).

small holes of $r \approx 1$ mm radius that does not allow microwaves to escape; the condition $\lambda >> r$ holds and, hence, it functions as a continuous metallic, highly reflecting, surface. The receptacle we see when we introduce the meal to be microwaved is the resonant cavity as shown in **Figure 3**. Electromagnetically, a rectangular resonant cavity, beneath the plastic covers, conforms to a metallic box of about $30 \times 32 \times 38\ \text{cm}^3$ dimensions. A 3D standing wave pattern is self-established due to the boundary conditions that have to be fulfilled at the walls, see **Figure 1(C)**.

Hence, maxima and minima and zeroes of the electric field $\left(\vec{E}\right)$ and magnetic field $\left(\vec{B}\right)$ appear at different locations, x, y, z, inside the cooking chamber. These microwaves bouncing continuously from the six walls bath constantly and heat our meal.

Can the microwaves, green wavy arrows in **Figure 3**, escape from the cooking chamber? Not in principle, unless some malalignment, or broken piece is there. Within the resonant cavity, the microwaves bounce back and forth from the metallic walls (yellow in **Figure 3(B)**) without any loss of electromagnetic energy.

Then, the microwaves hit the chicken at multiple points (blue-green wavy arrows), at the interface between food and air, reflection and refraction take place, and Snell's law and Fresnel equations have to be fulfilled [24–29]. Some microwaves are reflected (blue-green lines), and others are transmitted inside the chicken body (red wavy arrows). These red microwaves are responsible for heating, and they are the ones that transmit, quite efficiently, vertiginous motions vibrations, at 2.45 GHz, to the electric dipoles that are part of the meal (mostly water, but also some fatty molecules). These red microwaves penetrate several centimeters through the specimen. The electromagnetic energy carried out by the Poynting vector of these red microwaves, $\vec{S}(r,t) = \vec{E}(r,t) \times \vec{H}(r,t)$, is converted into frenetic Jiggling of these polar molecules and then converted into heat by their interactions with surrounding, neighboring molecules. Heat can be so high that some steam (water vapor) can be seen through the window in just a few seconds, see **Figure 3 (A)**. This process is the moment of energy conversion: from electromagnetic energy with 1000 Watts of power to motion, vibrations, mechanical energy. But that excited motion starts, rapidly, to pass to neighboring nonpolar atoms and molecules and locally, all the surrounding matter, starts jiggling more and more, which is heat. *Microwave energy has been transformed into heat inside our meal being microwaved*. In **Figure 3**, the wavy red lines represent the microwaves that get into the specimen and excite the electric dipoles within it. We show that at multiple points of incidence-reflection-transmission on the food-air interface, this bathing is by no means uniform since the microwaves distribute inhomogeneously inside the resonant cavity, see **Figure 1(C)**. This is the reason why the specimen is placed on a rotating plate, so some homogeneous heating is achieved.

Normally, it is expected that the ceramic (or the plastic), from which the cup for coffee or water is made, does not get heated while the liquid inside it. In order to get such result, it is necessary to minimize the composition of electric dipoles in the structure of the ceramics, glass, or plastic that makes the cup or the dish, effectively rendering this object transparent to the microwaves.

(B) The metallic waveguide: The microwave radiation from the source is immediately channeled through a horn-like metallic collector toward a rectangular waveguide through multiple reflections on its conducting metallic walls, and the radiation is guided almost without attenuation to the resonant cavity of the MWO. A rectangular waveguide is shown in **Figures 1(B)** and **4(C)**. The good conductor quality of the waveguide is the responsible for no-attenuation microwaves at the

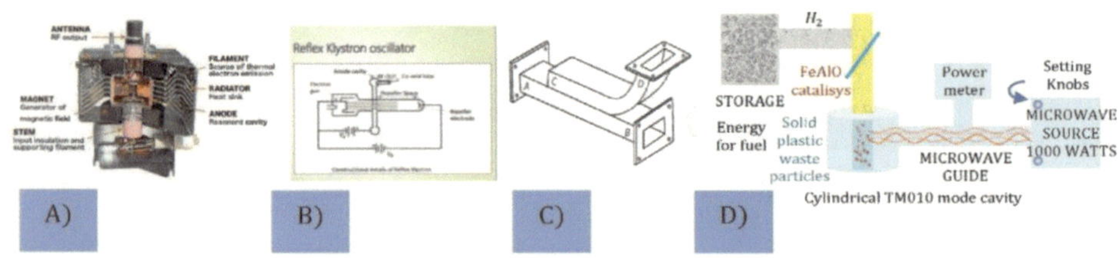

Figure 4.
Constituent parts of a microwave oven, or a laboratory microwave heater. (A) A magnetron typically used to generate microwaves in a domestic MWO. (B) A microwave generator, klystron, that is frequently used in microwave laboratory equipment where low power is required. (C) A hollow rectangular waveguide of a, b cross section and coupled to another waveguide, of the same dimensions, from aforesaid (taken from Feynman Phys. Lectures, vol. II. [24]) both microwave trains join and interfere at the union of the metallic structures. (D) An example of microwaving solids and trapping gasses that are detached, from the specimen, in the process. Microwaving plastic waste mixed 1:1 with FeAlOx inside the transformation chamber, which is an aluminum TM_{010} resonant cavity (it is the analog of the microwave oven chamber). Principle of operation was modified from [21], not the actual experimental setup.

waveguide walls even after multiple reflections. More detail on waveguides can be found in Refs. [24-29]. The waveguide terminates in a "mouth" that connects with the cooking chamber. This way both parts are coupled.

(C) The source of microwaves: For low-power applications, 1 mW or less, a klystron is used as the source of microwaves. For higher-power applications of microwave heating, around 1000 Watts, a magnetron is used. In both cases, it is a tube in which electrons are ejected from a hot cathode to a space where they get immediately accelerated and precisely, this acceleration produces electric and magnetic radiation fields, orthogonal to each other and to the propagation direction. Its power is proportional to the acceleration squared, Prad $\alpha\, a^2$, and is schematically depicted in **Figure 4(A)** and **(B)**. Those accelerated electrons emit radiation at the same frequency of the acceleration that in this case is in the 2–3 GHz range.

Now let us be more quantitative. We start with the electromagnetic boundary conditions.

3. Microwave heating of food or of an industrial sample

When microwave radiation hits the surface of a specimen, what we have is incidence of electromagnetic fields on the interface between meal and air, two nonmagnetic, nonconducting media, and the laws of Snell and Fresnel of reflection and refraction have to be fulfilled. But, they will be obeyed once the boundary conditions for $\vec{E}$ and $\vec{D}$ and for $\vec{H}$ and $\vec{B}$ are fulfilled. Let us see what they are as follows:

3.1 Continuity of the normal component of the displacement field, *D*

Let us suppose we have the interface between any two media such as water-air, plastic-metal, raw meat-hot air, a catalyst-plastic, ceramic-coffee, and so on as shown in **Figure 5**, that is, the boundary. The boundary condition on $\vec{D}$ is obtained from applying Gauss law to a very small cylinder (purple) of differential area and differential high that crosses the boundary as shown in **Figure 5(A)**. Then, the Gauss integral $\oint \vec{E} \cdot \vec{n} da$ on the closed surface is decomposed into three integrals, one on S_1 within medium 1, another one on surface S_2 within medium 2, and a third integral on the lateral surface, which goes to zero because the high of the cylinder is as small as we wish; then after integrating the only two integrals we are left with

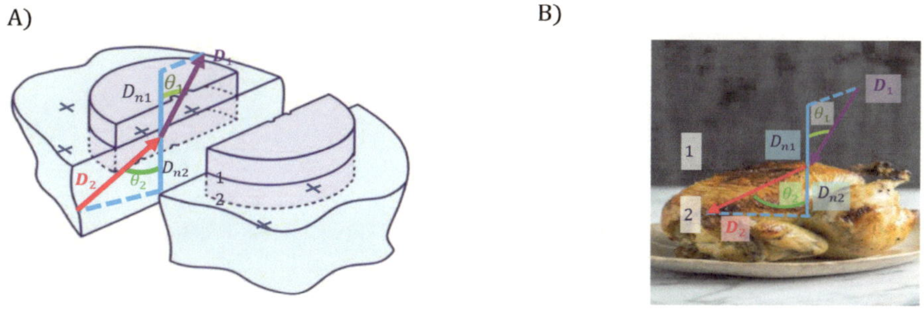

Figure 5.
Boundary conditions on $\vec{D}$. (A) Gaussian, very small, cylinder on the interface between two different media 1 and 2. The difference $D_{n1} - D_{n2}$ between the normal components of $\vec{D}$ is equal to the surface charge density σ_f. When surface charge is zero, then $D_{n1} = D_{n2}$, the normal components of the displacement field are continuous. (B) The same condition applies on the surface of meat when field $\vec{D}$ hits its surface inside a microwave oven, in this case $\sigma_f = 0$ and $D_{n1} = D_{n2}$. In (B) the boundary interface is the skin of a chicken being microwaved.

produce $(D_{n1} - D_{n2})S = \sigma_f S$ [25–29]. When the interface carries no electric charge, as is usually the case with microwave heating, then $\sigma_f = 0$. Hence, $(D_{n1} = D_{n2})$. And so, the normal component of the displacement vector is continuous through the interface of air-chicken!, or air-plastic, or air-ceramic, and so on.

Only in the case of the boundary between a conductor and a dielectric $D_n = \sigma_f$, being σ_f the free charge density on the interface as represented with the "+" signs in **Figure 5(A)**. For the cases we are interested here, all the metallic walls in the cooking chamber (resonant cavity) and the waveguide are not charged, then $\sigma_f = 0$ and so $D_{n1} = D_{n2}$.

On the other hand, when the $\vec{D}$, field of the microwaves in the cooking chamber enters the surface, D_{n2}, of the piece of food (sample, specimen, system), as shown in **Figure 5(B)**, it travels much more distance inside (several centimeters.) the sample and in its way excites electric dipoles (mostly water molecules and fat dipolar and other organic dipolar moieties) and gradually, but fast, transfers most of its energy to them. After a relaxation time period, $\approx 10^{-6}$ sec, the dipoles transfer all that juggling energy to vibrations of the bulk and appear as heat (measured as $k_B T$).

3.2 Continuity of the tangential component of the electric field intensity

Consider the blue, rectangular, path shown in **Figure 6(A)** and **(B)** with two sides parallel to the boundary and arbitrarily close to it. The two vertical sides are infinitesimal. Stokes theorem states that $\oint \left(\nabla \times \vec{E}\right) \cdot d\vec{a} = \oint \vec{E} \cdot d\vec{l}$. If the vertical paths are as short as we wish, E_t does not vary significantly over them and their integrals are zero. And the line integral of $\vec{E} \cdot d\vec{l}$ is $E_{t1}L–E_{t2}L$. By Stokes's Theorem, this line integral is equal to the integral of $\nabla \times \vec{E}$ over the surface enclosed by the path C [25–29].

By definition, the enclosed area is zero. So, $E_{t1}L–E_{t2}L = 0$, hence $E_{t1} = E_{t2}$. The tangential component of $\vec{E}$ is therefore continuous across the boundary. Applying this reasoning to the interface of a piece of food and the air in a microwave oven, as shown in **Figure 6(B)**, everything follows and the tangential component of $\vec{E}$ in the cooking chamber just above the surface of that matter is continuous with the tangential component of $\vec{E}$ just inside "the chicken."

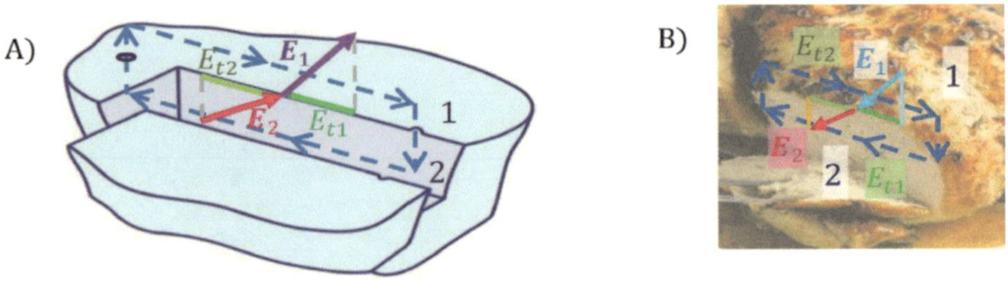

Figure 6.

Boundary condition on $\vec{E}$. (A) Closed path of integration crossing the interface between two different media 1 and 2. Whatever be the surface charge density σ_f, the tangential components of $\vec{E}$ on both sides of the interface are equal: $E_{t1} = E_{t2}$. The tangential components of $\vec{E}$ field are continuous no matter what medium 1 is and what medium 2 is. (B) The same analysis for a piece of chicken, or a cup of coffee, or melting cheese in a microwave oven follows: The tangential components of $\vec{E}$ are continuous. E_{t_2} contributes, at most, to some heating on the surface of the meal.

For the case of the metallic walls of the heating chamber and the walls of the waveguide, we have the case of a boundary between a dielectric (hot air) and a conductor, then $\vec{E} = 0$ in the conductor and $E_t = 0$ in both media. The magnetic part of the microwaves also obeys corresponding boundary conditions, namely: $B_{1n} = B_{2n}$ and $H_{1t} = H_{2t}$, and B_{2n} is quite capable of exciting magnetic dipoles inside the specimen, but food, beverages, water, and coffee do not possess magnetic moments, which are not magnetic. So, we do not treat here magnetic heating, even though it is a very active field of research [10, 11]. We concentrate on heating through electric dipoles inside the cooking chamber, as shown in **Figures 3, 5**, and **6**. Next, we describe more quantitatively the electromagnetics in the cooking chamber.

4. Microwave cooking chamber as a resonant cavity

Aforesaid, the empty cooking chamber in a microwave oven (MWO) is a closed rectangular space where, once microwaves are input, they bounce back and forth from metallic walls on the six sides and confine the electromagnetic waves in such space. This is an electromagnetic resonant cavity (ERC) in which electromagnetic waves (EMW) move in space and time periodically and, very importantly, forming standing wave patterns with nodes and anti-nodes. The cooking chamber is then electromagnetically a resonant cavity that imposes on the microwaves boundary conditions at the six walls. The $\vec{E}$ field, just outside and parallel to each wall, E_t, must be zero and the normal component of $\vec{B}$ must be continuous [25–29]. When food, water, coffee, cheese, or any other food are introduced in it, a dielectric medium with $\varepsilon \neq \varepsilon_0$ and air with $\approx \varepsilon_0$ are now the composite dielectric that fills the resonant cavity, as shown in **Figure 3**. Dielectrics do not perturb considerably the standing wave patterns that form the microwaves inside the MWO.

Let us be more quantitative, and take a standard microwave oven of dimensions $a = 30$ cm, $b = 32$ cm, and $c = 38$ cm as shown in **Figure 1(D)**. We will take the walls as perfect conductors as first approximation, boundary conditions on the six walls have to be fulfilled, and there will be multiple reflections at the metallic boundary surfaces. **Figure 7** shows how a sinusoidal electromagnetic field wave bounces back from a perfect conducting surface [25–29].

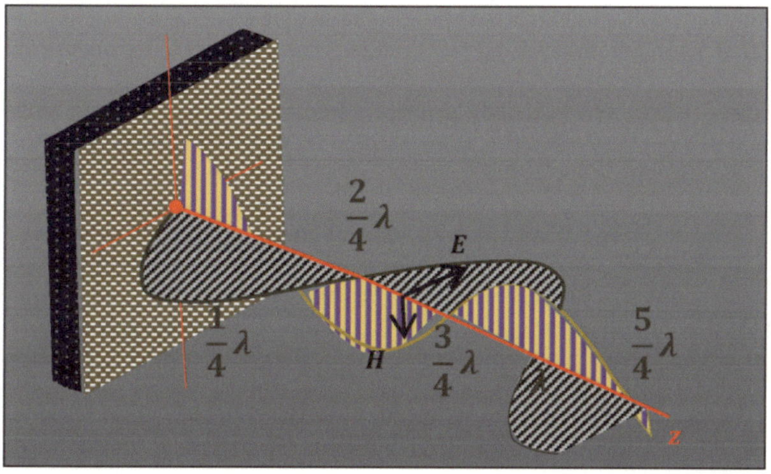

Figure 7.
The standing wave pattern resulting from the reflection of a microwave at the surface of a good conductor wall of the cooking chamber. The curvy lines show the standing waves of $\vec{E}$ and $\vec{H}$ at some particular time. The nodes $\vec{E}$ and of $\vec{H}$ are not coincident but are spaced λ/4 apart as shown. Modified from [25].

An arbitrary standing wave pattern in the resonant cavity can then be obtained as an appropriate superposition of these standing waves. Let us consider the closed region (cooking chamber) with walls of sides a, b, and c, and with the origin at one corner as shown in **Figure 3**. The cavity is filled with a linear dielectric, food (material described by μ_0 and ε). Both fields, $\vec{E}$ and $\vec{H}$, inside should obey Maxwell equations and each field component satisfies the wave equation; the solutions are stationary, confined, trapped microwaves. Applying separation of variables first to the time variable, it results in a solution of the form $e^{-i\omega t}$. Therefore, we have now, let us say, for

$$x\left(\vec{r},t\right) = \psi_0 = E_{0_x}X(x)Y(y)Z(z)e^{-i\omega t} \tag{1}$$

which, when substituted into the wave equation, leads to the Helmholtz equation $\nabla^2\psi_0 + k_0^2\psi_0 = 0$ where $k_0^2 = (\omega/v)^2$. This result is actually valid for any kind of coordinate system. Helmholtz equation is readily solved in rectangular coordinates by separation of variables. If we write, $\psi_0 = X(x)Y(y)Z(z)$ and proceed with the standard separations, we obtain [28, 29].

$$\begin{aligned}\Psi_0\left(\vec{r}\right) &= (\mathrm{C_1 \sin k_1x + C_2 \cos k_1x})(\mathrm{C_3 \sin k_2y + C_4 \cos k_2y}) \\ &\quad \times (\mathrm{C_5 \sin k_3z + C_6 \cos k_3z})\end{aligned} \tag{2}$$

where k_1, k_2, and k_3 are the wave numbers in the x, y, z dimensions and are related to the frequency ω of the microwave field by the dispersion relation $k_1^2 + k_2^2 + k_3^2 = k_0^2 = (\omega/v)^2$. Combining this $\psi_0\left(\vec{r}\right)$ with the temporal solution $T(t)$, we get any one of the components of the $\vec{E}$ and $\vec{H}$ field as

$$\begin{aligned}\mathrm{E_x} &= (\mathrm{C_1 \sin k_1x + C_2 \cos k_1x})(\mathrm{C_3 \sin k_2y + C_4 \cos k_2y}) \\ &\quad \times (\mathrm{C_5 \sin k_3z + C_6 \cos k_3z})\mathrm{e}^{-\mathrm{i}\omega\mathrm{t}}\end{aligned} \tag{3}$$

The boundary conditions that obey each $\vec{E}$ and $\vec{H}$ field component are going to make the difference of the fields through the fact that the constants, C_1, C_2, ..., C_6, of each component will take different values, including zero. The boundary condition $E_{1t} = E_{2t}$ at any metallic wall ($E_{metal} = 0$) makes for tangential components to be zero. Hence, E_x will be a tangential component and must therefore vanish at the faces $y = 0$ and b and $z = 0$ and c. We see that this requires that $C_4 = C_6 = 0$ and that $k_2 = n\pi/b$ and $k_3 = p\pi/c$ for $\sin(k_2 b) = 0$ and $\sin(k_3 c) = 0$, and where n and p are integers.

Then we have for E_x:

$$E_x = (C_1' \sin k_1 x + C_2' \cos k_1 x) \sin k_2 y \sin k_3 z e^{-i\omega t} \tag{4}$$

where $C_1' = C_1 C_3 C_5$ and $C_2' = C_2 C_3 C_5$. Repeating this whole procedure for E_y and its boundary conditions and for E_z and its boundary conditions we get

$$E_y = \sin k_1 x (C_3' \sin k_2 y + C_4' \cos k_2 y) \sin k_3 z e^{-i\omega t} \tag{5}$$

$$E_z = \sin k_1 x \sin k_2 y (C_5' \sin k_3 z + C_6' \cos k_3 z) e^{-i\omega t} \tag{6}$$

and $k_1 = m\pi/a$. Substituting the expressions of k_1, k_2, k_3 into the dispersion relation, we obtain all the possible frequencies of oscillation in this cooking chamber of a, b, c dimensions

$$\left(\frac{\omega}{v}\right)^2 = (k_1^2 + k_2^2 + k_3^2) = \pi^2\left[\left(\frac{m}{a}\right)^2 + \left(\frac{n}{b}\right)^2 + \left(\frac{p}{c}\right)^2\right] \tag{7}$$

Notice that $k_1 = k_m$, $k_2 = k_n$, and $k_3 = k_p$. Each set of integers (n, m, p) define a mode of $\vec{E}$ field. From this relation aforesaid, we see that frequency ω takes only particular values determined by m/a, n/b, and p/c. There are many combinations of (n, m, p) called modes and then corresponding k values and "mode" frequencies ω_{nmp}. Each ω_{nmp} is a mode of vibration of the electric field and of the $\vec{H}$ field (that we obtain below).

It is important to note that if any two of the integers m, n, p are zero, then the other corresponding two k_1, k_2, k_3 are zero, and from the expressions for E_x, E_y, E_z we see then that all three components of $\vec{E}$ are zero. Hence, as a consequence all components of $\vec{H}$ become zero since $\vec{H} = \xi\vec{k} \times \vec{E}$ and no standing wave pattern is sustained in the cooking chamber. We define the vector wave number $\vec{k}$ with components $k_1 = k_n$, $k_2 = k_m$, and $k_3 = k_p$. The vector electric field must satisfy Maxwell's equations and, in particular, the first Maxwell equation (Gauss Law in differential form) with $\rho_f = 0$. We must have $\varepsilon\nabla \cdot \vec{E} = 0$. To apply divergence we construct $\partial Ex/\partial x$, $\partial Ey/\partial y$, $\partial Ez/\partial z$ with the field components above, we obtain

$$\begin{aligned} &-(k_1 C_2' + k_2 C_4' + k_3 C_6') \sin k_1 \mathrm{x} \sin k_2 y \sin k_3 z \\ &+[(k_1 C_1' \cos \mathrm{k}_1 \mathrm{x} \sin k_2 y \sin k_3 z) + (k_2 C_3' \sin k_1 \mathrm{x} \cos k_2 y \sin k_3 z) \\ &+(k_3 C_3' \sin \mathrm{k}_1 \mathrm{x} \sin k_2 y \cos k_3 z)] = 0 \end{aligned} \tag{8}$$

The three terms on the left must sum up to zero. One way to have this zero is to ask for each term individually be zero, then we set $k_1 C'_2 + k_2 C'_4 + k_3 C'_6 = 0$ *and also* $C'_1 = C'_3 = C'_5 = 0$. The only surviving constants are C'_2, C'_4, and C'_6 and the resulting field components are now

$$E_x = (C_2' \cos \mathrm{k}_1 \mathrm{x}) \sin k_2 y \sin k_3 z e^{-i\omega t} \tag{9}$$

$$E_y = \sin k_1 x (C_4' \cos \mathrm{k}_2 y) \sin k_3 z e^{-i\omega t} \tag{10}$$

$$E_z = \sin k_1 x \sin k_2 y \left(C_6' \cos k_3 z \right) e^{-i\omega t} \tag{11}$$

The amplitude of each of these waves is C_2', C_4', and C_6'. So, let us rename them as $C_2' = E_1$, $C_4' = E_2$, and $C_6' = E_3$, we find that the last conditions can be written as [28, 29].

$$k_1 E_1 + k_2 E_2 + k_3 E_3 = \vec{k} \cdot \vec{E} = 0 \tag{12}$$

The expressions for the field components finally become

$$E_x(x, y, z; t) = E_1 \cos k_1 x \sin k_2 y \sin k_3 z e^{-i\omega t} \tag{13}$$

$$E_y(x, y, z; t) = E_2 \sin k_1 x \cos k_2 y \sin k_3 z e^{-i\omega t} \tag{14}$$

$$E_z(x, y, z; t) = E_3 \sin k_1 x \sin k_2 y \cos k_3 z e^{-i\omega t} \tag{15}$$

So that E_1, E_2, and E_3 are the amplitudes of the respective components.

We now show, in **Figure 8**, a 3D plot of the E_y component for the mode $n = 2$, $m = 4, p = 3$, and a 3D plot of the E_z component for the mode $n = 3, m = 5, p = 2$. It is immediately apparent that the number of maxima, minima, and nodes increases as the mode number (n, m, p) increases. Both plots show in the horizontal plane the projections of these maxima and minima. When thinking in the rectangular microwave cavity, this 2D plot represents the heating power at different spots at a z = constant plane (height in the microwave oven). This is just a very simplified picture of what the hot and cold spots are inside the 3D microwave chamber. The whole hot/cold distribution spots are the superpositions of many (n, m, p) electromagnetic standing wave patterns.

The color code used in **Figure 8(A)** and **(B)** represents maxima (red) and minima (blue) of the electric field of the microwaves. Since the energy delivered to

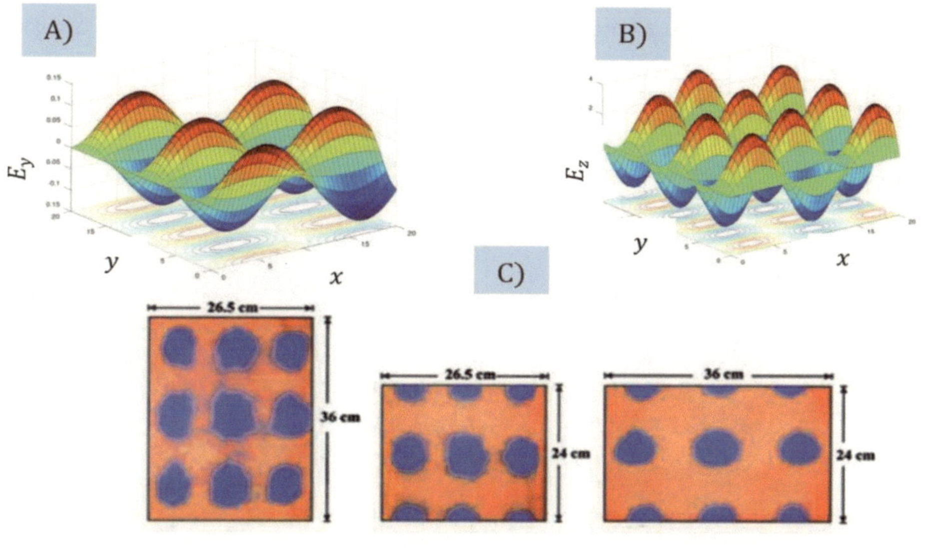

Figure 8.
Calculated and experimental electric field stationary wave patterns inside a rectangular cavity. (A) 3D plot of the E_y stationary pattern for the mode n = 2, m = 1, p = 3, evaluated from Eq. (14) at an arbitrary z fixed value. (B) 3D plot of the E_z stationary pattern for the mode n = 2, m = 1, p = 3, evaluated from Eq. (15) at an arbitrary z fixed value. In both cases, the projection in the x–y plane of the maxima, minima, and nodes is shown. (C) The experimental determination of the hot/cold spots in a rectangular chamber, a = 36 cm, b = 24 cm, and c = 26.5 cm. Note the alternating pattern of hot/cold spots in the stationary pattern (adapted from [30]).

the sample goes as the square of the electric field, then red and blue extrema become hot spots and the nodes become cold spots and are located between the red and blue spots. In the 2D projection, the regions between the blue and red zones are the cold spots. Experimental measurements on the standing wave patterns in microwave ovens have been reported and nicely agree with the theoretical calculations [30, 31]. In **Figure 8(C)**, we show one of those measurements carried out on three perpendicular planes, x–z, y–z, and x–y. The agreement between our calculations, planar 2D plots, and the experimental hot/cold spots in **Figure 8** is very satisfactory. The general pattern calculated and measured in the three planes consists of alternating maxima, minima, and nodes for each mode (n, m, p) as given by formulas (12)–(15).

The results in Eqs. (12)–(15) are very much like that one found for a plane wave [25–29]. Thus, for a given mode, a particular set (n, m, p), $\vec{E}_0$ must be perpendicular to the vector $\vec{k} = (m\pi/a)\hat{x} + (n\pi/b)\hat{y} + (p\pi/c)\hat{z}$.

For cylindrical cavities, stationary electromagnetic wave patterns are also obtained. Cylindrical cavities are very frequently used in research and in industrial applications as we showed above in **Figure 2(C)** for decomposition of waste plastics into H_2 and a set of fullerene solid compounds. Here, without any calculations, we show in **Figure 9** the electric field stationary wave pattern that we simulated from the cylindrical solutions for the TM_{010} mode. It is shown as a manner of contrast with the stationary pattern that results in rectangular geometry.

Calculating the magnetic field, we start from our knowledge of $\vec{E}$ and of the vector wave number $\vec{k}$. From the third Maxwell equation, we have $\nabla \times \vec{E} = -\mu\left(\frac{\partial \vec{H}}{\partial t}\right) = i\omega\mu\vec{H}$. For example, using the expressions for E_x, E_y, E_z just found above, we have

$$i\omega\mu H_x = \frac{\partial E_z}{\partial y} - \frac{\partial E_y}{\partial z} = (k_2E_3 - k_3E_2)\sin k_1 x \cos k_2 y \cos k_3 z e^{-i\omega t}. \quad (16)$$

Since k_2E_3–k_3E_2 is the x component of $\vec{k} \times \overrightarrow{E_0}$, it is desirable to define a vector $\overrightarrow{H_0}$ by $\overrightarrow{H_0} = \frac{1}{\omega\mu}\vec{k} \times \overrightarrow{E_0}$ for, if we let its rectangular components be H_1, H_2, and H_3, then we can write H_x as

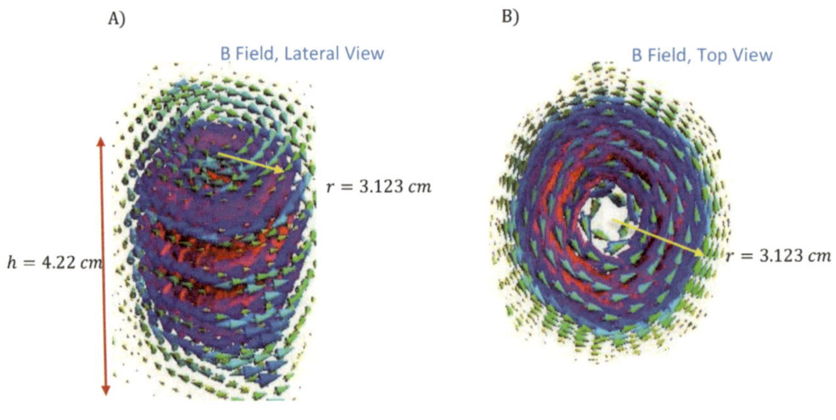

Figure 9.
Simulation of the stationary pattern of the electric field inside a cylindrical resonant cavity, TM_{010} mode. (A) The electric field is concentric with minima close to r = 0, and $\vec{E}$ = 0 exactly at r = 0. The field is tangent to the metallic wall and very small at R = r. (B) A top view of the same electric field stationary pattern [19]. This stationary field configuration is established in cylindrical cavities, TM_{010} mode, used for research at low microwave powers to excite magnetic specimens.

$$H_x = -iH_1 \sin(k_1 x)\cos(k_2 y)\cos(k_3 z)e^{-i\omega t} \tag{17}$$

Similarly, we find the other two components of $\vec{H}$ to be

$$H_y = -iH_2 \cos(k_1 x)\sin(k_2 y)\cos(k_3 z)e^{-i\omega t} \tag{18}$$

and

$$H_z = -iH_3 \cos(k_1 x)\cos(k_2 y)\sin(k_3 z)e^{-i\omega t} \tag{19}$$

We see that $H_x = 0$ at $x = 0$ and $x = a$, that is, at the walls for which it is a normal component; similarly, H_y and H_z vanish at $y = 0$ and b and $z = 0$ and c, respectively. Thus, the boundary conditions on $\vec{H}$ have been automatically satisfied once $\vec{E}$ was made to satisfy its own boundary conditions. Furthermore, it is easily verified that the two remaining Maxwell equations that we have not yet used are satisfied, that is, $\nabla \cdot \vec{H} = 0$ and $\nabla \times \vec{E} = -\left(\partial\vec{B}/\partial t\right)$ One needs to use $\vec{k} \cdot \overrightarrow{H_0} = 0$, as well as $\mathbf{H}_x E_z$ and the relation dispersion in its form $\vec{k} \cdot \vec{k} = k_0^2 = \frac{\omega^2}{v^2} = \omega^2\mu\varepsilon$. Each component of $\vec{E}$ varies as $e^{-i\omega t}$, while the components of $\vec{H}$ are proportional to $-ie^{-i\omega t} = e^{-i\left[\omega t + \left(\frac{1}{2}\right)\pi\right]}$. Thus, the electric and magnetic fields are not in phase in these standing waves but instead $\vec{H}$ leads $\vec{E}$ by 90° as shown in **Figure** 7. A given $\vec{k}$ corresponds to a given mode, that is, a given set of integers m, n, p in k aforesaid. Now $\vec{k} \cdot \vec{E}$ tell us that the vector E_0 must be chosen to be perpendicular to $\vec{k}$. However, there are two independent mutually perpendicular directions along which $\overrightarrow{E_0}$ can be chosen and still be perpendicular to a $\vec{k}$.

Thus, for each possible value of $\vec{k}$, there are two possible independent directions of polarization of $\overrightarrow{E_0}$, so that there are two distinct modes for each allowed frequency given by ω_{nmp}. This property is known as degeneracy and is a fundamental and important feature of electromagnetic standing waves. If a, b, c are all different, then the various frequencies given by ω_{nmp} will generally be different. However, if there are simple relations among the dimensions, it is possible that different choices of the integers will give the same frequency so that we will also have degeneracy, but arising in a different manner. As an extreme example, consider a cube for which $a = b = c$, so that ω_{nmp} reduces to $\left(\frac{\omega}{v}\right)^2 = \left(\frac{\pi}{a}\right)^2(m^2 + n^2 + p^2)$. Thus, all combinations of integers that have the same value of $m^2 + n^2 + p^2$ will have the same frequency and the modes will be degenerate.

5. The Poynting vector of the microwave fields inside the cooking chamber

Remembering that $\vec{S}(r,t) = \vec{E}(r,t) \times \vec{H}(r,t)$, taking the expressions of $\vec{E}$ and $\vec{H}$ Inside the cooking chamber, then we obtain $\vec{S}(r,t) = |E||H|\hat{k} = \xi \cdot v \cdot (\varepsilon/2)E^2\hat{k} = \xi \cdot vu\hat{k}$ [25–29]. This result is the general one obtained in any electrodynamic circumstance, of course, and microwave ovens fulfill it. And the average Poynting vector is $Sav = Power/Area = Energy/Area \cdot time$. What these expressions tell us is that microwave energy and microwave power inside the cooking chamber are traveling-moving, yes energy, and power, between the six walls in stationary wave patterns and in accord with the propagation vector $\vec{k}$ and carrying perpendicular to it the $\vec{E}$ and $\vec{H}$ fields. The microwave power deposited on a surface area of 1 mm^2 is then $Pav = Sav.A$. Since $\vec{E}$ and $\vec{H}$ inside the microwave oven have nodes and anti-nodes, then $\vec{S}(r,t)$ and consequently microwave power, $P(r,t)$ have nodes and anti-nodes at some positions along x, y, and z, and the heating is not uniform due to this standing wave feature of the microwave fields inside the cavity. Experimental results on the standing wave patterns have been reported and nicely agree with the theoretical calculations [30, 31]. Theoretically and experimentally standing microwave patterns are obtained. The reason of the rotating plate is to move in circular fashion the food to be heated and reach a more uniform microwave bathing on the food. Most of the time it is accomplished, but not always, as pizza fans report.

Now that we have a detailed treatment of the electromagnetic fields inside the cooking chamber, we want to develop some expressions for the $\vec{E}$ and $\vec{H}$ fields traveling on the waveguide from the magnetron toward the resonant cavity.

6. The waveguide in microwave heating systems, TE and TM modes

In research and technologically bound situations, the resonant cavities we saw above are feed with microwaves by means of waveguides connecting the source to the microwave cavity, see **Figures 2** and **4**. We can think of a waveguide as constructed from a cavity by taking the $\pm z$ walls to infinity; then, the trapped stationary waves in the cavity can now travel indefinitely toward $\pm$ infinity as plane waves. As soon as we start taking the $\pm z$ walls to infinity, we start liberating boundary conditions and in that dimension we are allowing free traveling waves. The remaining walls at $x = 0, a$, and $y = 0, b$ continue limiting our bouncing waves along these dimensions. We will continue taking the bounding surfaces as perfect conductors. A question to ask at this point is; Is it possible to transfer electromagnetic energy along a *waveguide*, that is, a tube with open ends? From everyday experience, we already know that this is possible from the simple fact that we can see through long straight pipes. So, the answer is yes, but: How is this carried out? Solutions to the wave equations have the answer, but first we review quickly boundary conditions on perfect conductors.

7. Boundary conditions at the surface of a perfect conductor

We recall that a *perfect conductor* is one for which $\sigma \to \infty$, more precisely, one for which the ratio $Q = \varepsilon\omega/\sigma \to 0$. $Q \leq (1/50) \ll 1$ for common metals even at very high frequencies so that $Q = 0$ should be a good first approximation for metallic boundaries. Plane waves traveling freely along the z direction take the form $E(x,y)\ \exp\ [i(\omega t - kz)]$, where the $E(x,y)$ part has to be found but we already know that fulfills boundary conditions at the metallic walls. We remember that $\delta = (2/\mu\sigma\omega)^{1/2}$ for a good conductor so that $\delta \to 0$ as $\sigma \to \infty$. Therefore, the electric field is zero at any point in a perfect conductor since the skin depth is zero. Since the tangential components of $\vec{E}$ are always continuous, we see that $\vec{E}_{tang} = 0$ just outside of the surface. In other words, $\vec{E}$ has no tangential component at the surface of a perfect conductor so that $\vec{E}$ must be normal to the surface [25–29]. $\vec{B}$ inside the conductor is $\vec{B} = (k/\omega)\vec{k} \times \vec{E_\tau}$ so that $\vec{B}$ will also be transverse. Consequently, the transverse component of $\vec{B}$ inside will also vanish as $\sigma \to \infty$. Since $\vec{B}$ has no normal component, the boundary condition $B_{1n} = B_{2n}$ implies that $\vec{B}_{norm} = 0$ just outside the conductor. Thus, at the surface of a perfect conductor and outside of it, $\vec{B}$ has no normal component; that is, it must be tangential to the surface. We see that *all* of the field vectors will be zero inside a perfect conductor. This simplifies greatly the general boundary conditions. To repeat: At the surface of a perfect conductor, $\vec{E}$ is normal to the surface and $\vec{B}$ is tangential to the surface. To put it another way, $\vec{E}$ has no tangential component while $\vec{B}$ has no normal component.

8. Propagation characteristics of waveguides

Figure 10 shows a waveguide that extends indefinitely in the z direction and of arbitrary and constant cross section in the xy plane. We take the boundary walls as perfect conductors and the interior of the cavity is filled with a linear nonconducting medium described by μ_0 and ε. If ψ is any component of $\vec{E}$ or $\vec{B}$, we know that it satisfies the scalar wave equation $\nabla^2\psi - \frac{1}{v^2}\frac{\partial^2\psi}{\partial t^2} = 0$ where $v^2 = 1/\mu\varepsilon$ and v would be the speed of a plane wave in the medium. Again, by separation of variables we easily find $\psi(x, y, z, t) = \psi_0(x, y)e^{i(k_g z - \omega t)}$.

We note that this is *not* a plane wave since the amplitude ψ_0 is not a constant but depends on x and y, the cross section [28, 29]. The quantity k_g is the *guide propagation constant, or simply the k_z constant of separation of the Z(z) component* of the whole solution and can be written as $k_g = 2\pi/\lambda_g$ here λ_g is the guide wavelength, that is, the spatial period along the guide, the z axis.

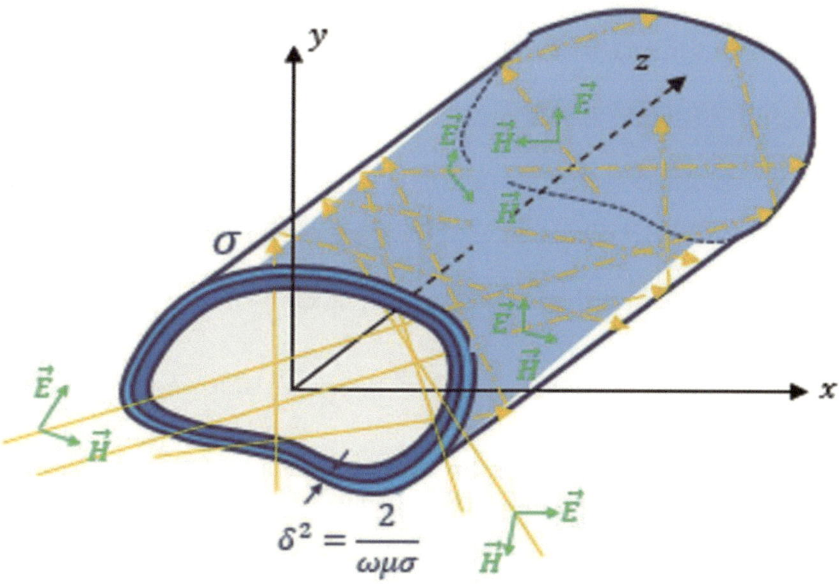

Figure 10.
A waveguide made of a perfect conductor with arbitrary and constant cross section. A set of propagation vectors k_1, k_2, k_3, *etc., are shown to impinge on different points on the metallic walls and reflect back following Snell law. Transmission is not depicted since perfect conducting walls are considered, and hence, the skin depth tends to zero, which implies zero transmission.*

Continuing with the separation of variables now for the x and y variables, we obtain again a Helmholtz equation in 2D $\frac{\partial^2\psi_0}{\partial x^2}+\frac{\partial^2\psi_0}{\partial y^2}+k_c^2\psi_0=0$, where $k_c^2=k_0^2-k_g^2$ and $k_0=\frac{\omega}{v}=\frac{2\pi}{\lambda_0}$. Writing $k_c=\frac{2\pi}{\lambda_c}$ we obtain a wavelength relation $\frac{1}{\lambda_c^2}=\frac{1}{\lambda_0^2}-\frac{1}{\lambda_g^2}$. Therefore, we have found for a waveguide that we will get wave propagation *only* if $k_0>k_c$, or $\lambda_0<\lambda_c$. For this reason, λ_c is called the *cutoff wavelength*. It is very common to state this result in terms of a cutoff frequency ωc defined by $k_c=\frac{\omega_c}{v}$ so that kg^2 can also be written as $k_g^2=\frac{1}{v^2}\left(\omega^2-\omega_c^2\right)$. Then, wave propagation is possible only if $\omega>\omega^2$, that is, if the applied frequency is greater than the cutoff frequency.

9. Rectangular guide

This guide has a rectangular cross section of sides a and b, which we take to be located in the xy plane. It is relevant to mention that for either type of mode, TE or TM, we have to solve an Helmholtz equation and apply boundary conditions as aforesaid. We continue using separation of variables and write $\psi_0(x,y)=X(x)Y(y)$; then, the same arguments as used in resonant cavity section above lead us to the separated equations

$$\frac{1}{X}\frac{\partial^2 X}{\partial x^2}=-\frac{1}{Y}\frac{\partial^2 Y}{\partial y^2}-k_c^2=\text{const.}=-k_1^2 \tag{20}$$

so that, $(d^2X/d\,x^2)+k_1^2X=0$ and $(d^2Y/d\,y^2)+k_2^2Y=0$ the separation constants have been selected with minus sign since the solutions should be periodic. Hence, $k_1^2+k_2^2=k_c^2$ is the dispersion relation in terms of the constants of separation for this 2D differential equation. Solving these in terms of sine and cosine functions, we find that $\psi_0(x,y)=(C_1\sin k_1x+C_2\cos k_1x)(C_3\sin k_2y+C_4\cos k_2y)$, where the C's are constants of integration. This expression for ψ_0 contains a total of four constants.

Let us calculate for TE *modes*. Here, we set $\mathcal{E}_z=0$ and write

$$\mathcal{H}_z=(C_1\sin k_1x+C_2\cos k_1x)(C_3\sin k_2y+C_4\cos k_2y) \tag{21}$$

With $\mathcal{E}_z=0$, and with $\nabla\times\vec{E}=-\partial\vec{B}/\partial t$, we find that when we substitute $\mathcal{H}_z$ into E_x and E_y coming from $\nabla\times\vec{E}$ and after some algebra [28].

$$\mathcal{E}_x=\frac{i\omega\mu k_2}{k_c^2}(C_1\sin k_1x+C_2\cos k_1x)(C_3\sin k_2y-C_4\cos k_2y) \tag{22}$$

$$\mathcal{E}_y=\frac{i\omega\mu k_1}{k_c^2}(C_1\cos k_1x-C_2\sin k_1x)(C_3\sin k_2y+C_4\cos k_2y) \tag{23}$$

From the boundary conditions $\mathcal{E}_x(y=0)=0$ and $\mathcal{E}_x(y=b)=0$ and similarly, from the boundary conditions that $\mathcal{E}_y(\mathrm{x})=0$ must satisfy at $x=0$ and $x=a$, then evaluating first for the zero values of x and y, we get

$$\mathcal{E}_x(x,0)=0=\frac{i\omega\mu k_2C_3}{k_c^2}(C_1\sin k_1x+C_2\cos k_1x) \tag{24}$$

$$\mathcal{E}_y(0,y)=0=-\frac{i\omega\mu k_1C_1}{k_c^2}(C_3\sin k_2y+C_4\cos k_2y) \tag{25}$$

Notice that we have here established a 2D homogeneous Sturm-Liuville problem and we expect to obtain as solutions eigenvalues and eigenfunctions. From the aforesaid boundary conditions, we must have $C_1 = 0$ and $C_3 = 0$. Therefore, at this stage, $\mathcal{E}_x$, $\mathcal{E}_y$, $\mathcal{H}_z$ have simplified to

$$\mathcal{H}_z = C_2 C_4 \cos k_1 x \cos k_2 y \tag{26}$$

$$\mathcal{E}_x = -\frac{i\omega\mu k_2 C_3}{k_c^2} C_2 C_4 \cos k_1 x \sin k_2 y \tag{27}$$

$$\mathcal{E}_y = \frac{i\omega\mu k_1 C_1}{k_c^2} C_2 C_4 \sin k_1 x \cos k_2 y \tag{28}$$

We still have boundary conditions to satisfy at the two remaining faces. We see that the requirement $\mathcal{E}_x(x, b) = 0$ leads to $\sin k_2 b = 0$ so that $k_2 b = n\pi$ where n is an integer. Similarly, $\mathcal{E}_y(\mathrm{a}, y) = 0$ gives the condition that $k_1 a = m\pi$ with m an integer. Thus, we have found the eigenvalues $k_1 = \frac{m\pi}{a}$ and $k_2 = \frac{n\pi}{b}$, these are the eigenvalues of the solution. So that $k_1^2 + k_2^2 = k_c^2$ shows that the allowed values of ${k_c}^2$ are $k_c^2 = k_{c\ mn}^2 \pi^2 \left[\left(\frac{m}{a}\right)^2 + \left(\frac{n}{b}\right)^2\right]$. The cutoff wavelengths and frequencies can now be found by using ${k_c}^2$ above into our ${\lambda_c}^2$ and ${\omega_c}^2$ equations. The corresponding guide propagation constants are

$$k_g^2 = \left(\frac{2\pi}{\lambda_g}\right)^2 = k_0^2 - \pi^2\left[\left(\frac{m}{a}\right)^2 + \left(\frac{n}{b}\right)^2\right] \tag{29}$$

The only quantity left undetermined is the arbitrary amplitude C_2C_4 of $\mathcal{H}_z$. If we set $C_2C_4 = H_0$, then we find that the amplitudes of a general TE mode in a rectangular guide are as follows:

$\mathcal{E}_x = -\frac{i\omega\mu}{k_c^2}\left(\frac{n\pi}{b}\right)H_0 \cos\left(\frac{m\pi x}{a}\right) \sin\left(\frac{n\pi y}{b}\right)$	$\mathcal{H}_x = -\frac{ik_g}{k_c^2}\left(\frac{m\pi}{a}\right)H_0 \sin\left(\frac{m\pi x}{a}\right) \cos\left(\frac{n\pi y}{b}\right)$
$\mathcal{E}_y = \frac{i\omega\mu}{k_c^2}\left(\frac{m\pi}{a}\right)H_0 \sin\left(\frac{m\pi x}{a}\right) \cos\left(\frac{n\pi y}{b}\right)$	$\mathcal{H}_y = -\frac{ik_g}{k_c^2}\left(\frac{n\pi}{b}\right)H_0 \cos\left(\frac{m\pi x}{a}\right) \sin\left(\frac{n\pi y}{b}\right)$
$\mathcal{E}_z = 0$	$\mathcal{H}_z = H_0 \cos\left(\frac{m\pi x}{a}\right) \sin\left(\frac{n\pi y}{b}\right)$

where k_c and k_g are as above. Multiplying each of these amplitude factors by the wave propagation term we get, for example, for the E_x field

$$E_x = -\frac{i\omega\mu}{k_c^2}\left(\frac{n\pi}{b}\right)H_0 \cos\left(\frac{m\pi x}{a}\right) \sin\left(\frac{n\pi y}{b}\right)e^{i(k_g z - \omega t)} \tag{30}$$

Since $-i = e^{-i(1/2)\pi}$, the exponential factor can be written $\exp i\left[k_g z - \left(\omega t + \frac{1}{2}\pi\right)\right]$, which shows that E_x leads H_z in time by 90°. Similarly, H_x and H_y lead H_z by 90° while E_y lag H_z by this same amount.

We now particularize to the simplest case which is also the most used. The TE_{10} *mode*, we set m = 1 and $n = 0$ and we particularize the above equations for these particular values of m and n, we show now without calculations that:

$$k_g = \left[\left(\frac{\omega}{v}\right)^2 - \left(\frac{\pi}{a}\right)^2\right]^{1/2}.$$

The field amplitudes are

$$\mathcal{E}_y = i\omega\mu\left(\frac{a}{\pi}\right)H_0 \sin\left(\frac{\pi x}{a}\right) \tag{31}$$

$$\mathcal{H}_x = -ik_g\left(\frac{a}{\pi}\right)H_0 \sin\left(\frac{\pi x}{a}\right) \tag{32}$$

$$\mathcal{H}_z = H_0 \cos\left(\frac{\pi x}{a}\right) \tag{33}$$

While $\mathcal{E}_x = \mathcal{E}_z = 0$ and $\mathcal{H}_y = 0$. Inserting these amplitudes into the complete $\vec{E}$ and $\vec{H}$ field expressions above and taking the real parts of the resulting expressions, we find the only nonzero field components to be

$$E_y = -H_0\omega\mu\left(\frac{a}{\pi}\right) \sin\left(\frac{\pi x}{a}\right) \sin\left(k_g z - \omega t\right) \tag{34}$$

$$H_x = H_0 k_g\left(\frac{a}{\pi}\right) \sin\left(\frac{\pi x}{a}\right) \sin\left(k_g z - \omega t\right) \tag{35}$$

$$H_z = H_0 \cos\left(\frac{\pi x}{a}\right) \cos\left(k_g z - \omega t\right) \tag{36}$$

We see that the values of the E_y are independent of y; hence, the electric field lines are straight lines with constant magnitude at a given value of x but with a magnitude that does vary with x and is a maximum at the center where $x = (a/2)$. The lines of H_x are straight with their maximum value at the center as well. The value of H_z, on the other hand, is zero in the center as has opposite signs on the two sides of the center.

TM *modes*. In this case, we set $\mathcal{H}_z = 0$ and set $\mathcal{E}_z$ equal to the expression for ψ_0 given above; hence, $\nabla \times \vec{E} = -\partial\vec{B}/\partial t$ is again applicable. This case is actually simpler because $\mathcal{E}_z$ can be a tangential component and must vanish for $x = 0$ and a and $y =$ and b. We have again an homogeneous Sturm-Liouville problem and expect eigenvalues and eigenfunctions. Proceeding in the same manner as before, we see that we must now have $C_2 = C_4 = 0$, while k_1, k_2, and k_c are given, again as above. Thus, the TE and TM modes of a rectangular waveguide have the same set of cutoff wavelengths, the same eigenfrequencies, and cutoff frequencies; the field configurations can be expected to be different however. Setting $C_1C_3 = E_0$, we find that (21) gives the starting point for the TM calculation to be $\mathcal{E}_z = E_0 \sin\left(\frac{m\pi x}{a}\right) \sin\left(\frac{n\pi y}{b}\right)$. We now use this $\mathcal{E}_z$ to calculate the rest of the field amplitudes following the above procedure. We note that $m = n = 0$ makes $\mathcal{E}_z$ and then all the other field components, zero; thus, there is no TM_{00} mode. Furthermore, if $m = 0$ or $n = 0$, $\mathcal{E}_z = 0$ and all of the fields are zero. Thus, it is not possible to have a TM_{m0} or TM_{0n} mode, in contrast to the TE case.

We have now calculated in detail the electric and magnetic fields that propagate in rectangular waveguides as the ones shown in **Figures 1(B)**, **2(A)**, and **3(C)**. The field patterns are stationary wave patterns in the x–y direction and traveling waves along the z direction as given by Eqs. (30)–(36).

Let us proceed now to the last of the physical components of a microwave heater, the very source of 1000 Watts microwaves.

10. Radiation of accelerated point charges in klystrons and magnetrons

Klystrons and magnetrons produce microwaves that carry power; typically, klystrons are used when little power is needed, from 1 watt to milliwatts and even microwatts. Magnetrons are used in higher-power applications, 1000 Watts, or more. Clearly, magnetrons are better suited for microwave heating. Both rely on electrons being accelerated (these electrons are labeled e_m) and made move in periodic trajectories inside cylindrical chambers. Both devices are shown in **Figure 11**. Notice the motion of electrons in them. Straight trajectories in klystrons, $w(t)$, **Figure 11(A)**, and curved trajectories, $Ç(t)$, in magnetrons, see **Figure 11(B)**. To expose relevant physics of accelerated electrons, em, produced in klystrons and Magnetrons is to describe the $\vec{E}_{rad}$ and $\vec{B}_{rad}$ microwave radiation fields produced inside their structures and from these, the Poynting vector, $\vec{S}\left(\vec{r},t\right) = \vec{E}_{rad} \times \vec{H}_{rad}$ that describes flux of such energy through space.

A microwave power, $P_{rad} = \left|\vec{S}\right| \cdot Area$, comes with this traveling energy. For klystrons, the accelerated electrons travel along straight lines, inside vacuum tubes, back and forth, due to voltage differences, $\Delta V_{12} \geq 860$ Volts, applied at the ends of the cylindrical tube (chamber), see **Figure 11(A)**. So acceleration is linear and is $\vec{a}(z,t) = {d\vec{v}(z)}/_{dt} = \frac{d^2\vec{w}(z,t)}{dt^2} = \vec{\ddot{w}}(z,t)$, in which $\vec{a} = \vec{\ddot{w}}$ is parallel to $\vec{v}$ and parallel to the tube axis, $\vec{z}$, see **Figure 11(A)**, and the acting force producing such acceleration is $\vec{F} = e\vec{E} = e(-\nabla V_{12})$. For magnetrons, the trajectories of the accelerated electrons are wavy circular with average radius $a \leq r \leq b$, as shown in **Figure 11(B)** and **(C)**. The wavy $\vec{ç}\left(\vec{r},t\right)$ trajectories of the accelerated electrons in magnetrons are the effect of a combined magnetic force $\vec{F}_{mag} = q\dot{\vec{ç}}(t)xB\hat{e}_z$ and the total electric force between these electrons and the charges located in pairs along the b radius, $\sigma\pm$ and $-Q$ at the cathode. What we have now is, charged particles, e_m, moving along trajectories, $\vec{w}(z,t)$, in klystrons, and $\vec{ç}(z,t)$, in magnetrons, both have velocities $\dot{\vec{w}}(z,t)$, $\dot{\vec{ç}}(z,t)$, and accelerations, $\vec{\ddot{w}}(z,t)$, $\ddot{\vec{ç}}(z,t)$. Charged particles in motion produce electric potentials and electromagnetic fields just as static charges do,

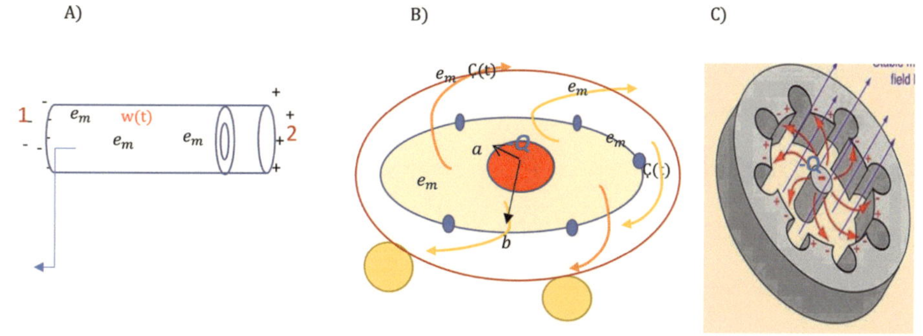

Figure 11.
Microwave sources, reflex klystron, and magnetron. (A) The basics of a klystron that produces accelerated electrons through an alternating electric potential difference $\Delta V_{12}(\omega t)$, these e_m travel the distance d; then, the acceleration is reversed, e_m travel to the left now, and this repeats thousands of times at a GHz frequency. (B) In a magnetron hot electrons ejected from a central cathode, travel in circular-wavy trajectories inside a cylinder due to the Lorentz force $\vec{F} = e\left(\vec{E} + \vec{v}(t)xB\hat{e}_z\right)$, where $\vec{v}(t) = \partial\vec{ç}\left(\vec{r},t\right)/\partial t = \dot{\vec{ç}}\left(\vec{r},t\right)$ is the velocity of the e_m electrons, $\vec{E}$ is the total electric field due to the perimetral charges (+, −), (+, −), (+, −), and the central −Q charge. $\vec{B}$ is a constant magnetic field (from a magnet) applied along the $\hat{e}_z$ axis. These two forces combined produce the curved-wave trajectory $\vec{ç}\left(\vec{r},t\right)$. (C) A complete diagram of the magnetron structure with the constant magnetic field $B\hat{e}_z$, the charge distribution (proper of magnetrons) that produces a total $\vec{E}$ field and the curved electron trajectories $\vec{ç}\left(\vec{r},t\right)$.

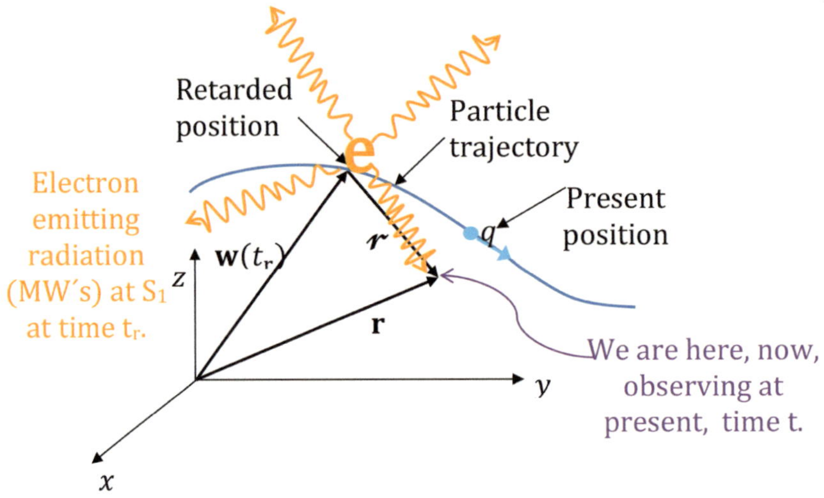

Figure 12.
A moving charged particle following a trajectory w(tr). $|\vec{r}\text{-}\vec{w}(t_r)|$ *is the distance the "radiated field" from the moving electron must travel, and* $(t - t_r)$ *is the time it takes to make the trip, we shall call* $\vec{w}(t_r)$ *the* ***retarded position*** *of the charge,* $\vec{\mathscr{r}}$ *is the vector distance from the retarded position to the point the radiated electromagnetic wave arrived "to us" (now, at our time* t_2*), which is at* $\vec{r}$*, clearly* $\vec{\mathscr{r}} = \vec{r}\text{-}\vec{w}\ (t_r)$.

except that here we have to calculate retarded potentials and retarded fields. For single charged particles, the resulting potentials are the well-known Lienard-Wiechert potentials [26–29, 32]. We present them here now. We take an accelerated electron moving in a general trajectory given by $\vec{\varsigma}\left(\vec{r},t\right)$, or $w(r,t) \equiv$ position of q at time t. Now, $V\left(\vec{r},t\right) = \frac{1}{4\pi\epsilon_0}\int\frac{\rho\left(\vec{r}',t_r\right)}{\mathscr{r}}d\tau'$ gives the electric potential at r at time t (**Figure 12**) [2, 26–29].

The retarded integration is not trivial, and the retardation in the $q/\mathscr{r}$ term in the potential aforesaid throws in a factor $\frac{q}{1-\vec{\mathscr{r}}\cdot\vec{v}/c}$, where $\vec{v}$ is the velocity of the charge at the retarded time, and $\vec{\mathscr{r}}$ is the vector from the retarded position to the point $\vec{r}$ where we are standing and measuring [24, 26–29, 32]. Then, $\int\rho\left(\vec{r}',t_r\right)d\tau' = \frac{q}{1-\vec{\mathscr{r}}\cdot\vec{v}/c}$. It follows, then, that $V\left(\vec{r},t\right) = \frac{1}{4\pi\epsilon_0}\frac{qc}{\left(\mathscr{r}c-\vec{\mathscr{r}}\cdot\vec{v}\right)}$. Meanwhile, since the current density of a rigid object is $\vec{J} = \rho\vec{v}$, we also have $\vec{A}\left(\vec{r},t\right) = \frac{\mu_0}{4\pi}\int\frac{\rho\left(\vec{r}',t_r\right)\vec{v}(t_r)}{\mathscr{r}}d\tau' = \frac{\mu_0}{4\pi}\frac{\vec{v}}{\mathscr{r}}\int\rho\left(\vec{r}',t_r\right)d\tau'$.

Or $\vec{A}\left(\vec{r},t\right) = \frac{\mu_0}{4\pi}\frac{qc\vec{v}}{\mathscr{r}c-\vec{\mathscr{r}}\cdot\vec{v}} = \frac{\vec{v}}{c^2}V\left(\vec{r},t\right)$. These are the famous Lienard-Wiechert potentials for a moving point charge. By using $\vec{E} = -\nabla V - \partial\vec{A}/\partial t$ and $\vec{B} = \nabla\times\vec{A}$, the corresponding fields are evaluated. It seems to us that only two authors, Jefimenko and Griffiths, give detailed derivation of these fields. The differential operations should be carried out with great care as these authors do, and we refer to those calculations and just take their results here:

$$\vec{E}\left(\vec{r},t\right) = \frac{q}{4\pi\epsilon_0}\frac{\mathscr{r}}{\left(\vec{\mathscr{r}}\cdot\vec{u}\right)^3}\left[(c^2-v^3)\vec{u}+\vec{\mathscr{r}}\times\left(\vec{u}\times\vec{a}\right)\right] = \vec{E}_{vel}+\vec{E}_{accl} \tag{37}$$

where $\vec{u} = c\hat{\mathscr{r}} - \vec{v}$, and, very importantly, $\vec{E}_{vel}\ \alpha\ 1/\mathscr{r}^2$ and $\vec{E}_{accel}\ \alpha\ \mathscr{r}$. And the magnetic field is

$$\vec{B}\left(\vec{r},t\right)=\frac{1}{c}\vec{\boldsymbol{r}}\times\vec{E}\left(\vec{r},t\right)\tag{38}$$

$\vec{B}$ follows the same time dependence as $\vec{E}$. The first term in $\vec{E}\ (\vec{r},t)$ is called the **velocity field**, and the triple cross-product is called the **acceleration field**. With these potentials and Jefimenko fields, we are now in the position to describe more quantitatively the radiation produced in klystrons and magnetrons. Hence, the accelerated electrons, e_m, inside these devices produce V and $\vec{A}$ potentials and $\vec{E}$ and $\vec{B}$ fields. In klystrons, the radiation fields are, as they are, captured by the mouth of a waveguide and send to a resonant cavity in which they produce standing patterns of stationary microwaves for their use.

Finally, in magnetrons, the perimetral charges, $\sigma\pm$ experience Lorentz forces due to these Jefimenko fields, $\vec{F}=q^{\mp}\left(\vec{E_J}+\dot{\vec{\varsigma}}(t)x\vec{B_J}\right)$; hence, these charges move inside the conducting core behind radius b and around the cylindrical cavities, see **Figure 9(C)**, and cut from the solid metal (usually copper). These moving charges, in turn, produce their own retarded potentials, $V_\sigma(wt)$, $\vec{A}_\sigma(wt)$ and fields, $\vec{E}_\sigma(wt)$ and $\vec{B}_\sigma(wt)$. We end up with total fields, $\vec{E}_t(wt)$ and $\vec{B}_t(wt)$, inside the magnetron space, including the cylindrical cavities (eight of them most of the time) behind the radius b. These cylindrical cavities are there, precisely, to trap microwaves in them and due to their perfect conducting walls, $\vec{E}_{tot}(wt)$ and $\vec{B}_{tot}(wt)$ reflected from them with almost no losses, and so these cavities sustain stationary microwave patterns of cylindrical geometry. The same process takes place in the eight cylindrical cavities distributed along the perimeter of radius b. With a simple wire antenna, microwaves are taken out of these cavities and sent to the entrance of the waveguide; then, these microwaves travel the short distance inside the waveguide and end up in the cooking chamber of the microwave oven; hence, our coffee absorbs so much the energy of these microwaves; the electric dipoles in the water vibrate and jiggle; frenetically, at 2.45 GHz, in a few seconds our coffee is hot and ready to drink.

11. Conclusions

In this chapter, detailed electrodynamic descriptions of the fundamental workings of microwave heating devices were given. We analyzed one by one the principal components of a microwave heater; the cooking chamber, the waveguide, and the microwave sources, either klystron or magnetron. The boundary conditions at the walls of the resonant cavity and at the interface between air and the surface of the food were stressed. It was shown how relevant the boundary conditions are to understand how the microwaves penetrate the nonconducting, electric polarizable specimen. In addition to microwave food, we mentioned the important application of microwaving waste plastics to obtain a good H_2 quantity that could be used as a clean energy source for other machines and so contributing to a cleaner planet. We did use Maxwell equations to obtain trapped stationary microwaves in the resonant cavity and traveling waves in the waveguides. We showed 3D plots of a few lower E_x, E_y, E_z modes calculated directly from the solutions obtained here and compared the general trend with experimentally obtained microwave heated patters inside rectangular cavities. The agreement is very good. We did simulate a single electro-magnetic field mode inside a cylindrical cavity in order to contrast with the stationary patterns obtained in rectangular cavities. The radiation processes in klystrons and magnetrons were stated in terms of the accelerated electrons produced. Then, using the Lienard-Wiechert potentials produced by these electrons, the Jefimenko fields were written. When all these are put together, we understand how a meal or a waste plastic, or an industrial sample, is microwave heated.

Author details

Rafael Zamorano Ulloa
Superior School of Physics and Mathematics, IPN, Mexico City, Mexico

*Address all correspondence to: davozam@yahoo.com

References

[1] Penzias AA, Wilson RW. A measurement of excess antenna temperature at 4080 Mc/s. Astrophysical Journal Letters. 1965;**142**: 419-421. DOI: 10.1086/148307

[2] Microwave Ablation 'Safe and Effective Technique' at Treating Large Benign Thyroid Nodules [Internet]. 2020. Available from: https://www.healio.com/news/endocrinology/20201020/microwave-ablation-safe-and-effective-technique-at-treating-large-benign-thyroid-nodules [Accessed: 14 March 2021]

[3] Cancer Research. Microwave Ablation. Lung Cancer [Internet]. UK: Cancer Research; 2021. Available from: https://www.cancerresearchuk.org/about-cancer/lung-cancer/treatment/microwave-ablation [Accessed: 14 March 2021]

[4] Makul N, Rattanadecho P, Pichaicherd A. Accelerated microwave curing of concrete: A design and performance related experiments. Cement and Concrete Composites. 2017; **i3**:415. DOI: 10.1016/j.cemconcomp.2017.08.007

[5] Halim SA, Swithenbank J. Simulation study of parameters influencing microwave heating of biomass. Journal of the Energy Institute. 2019;**92**:1191-1212. DOI: 10.1016/j.joei.2018.05.010

[6] Mingos M, Baghurst D. Applications of microwave dielectric heating effects to synthetic problems in chemistry. Chemical Society Reviews. 1991;**20**:1-47. DOI: 10.1039/CS9912000001

[7] Kappe O, Dallinger D. The impact of microwave synthesis on drug discovery. Nature Reviews. Drug Discovery. 2006; **54**:5. DOI: 10.1038/nrd1926

[8] Vollmer M. Physics of the microwave oven. Physics Education. 2004;**39**:74-81. DOI: 10.1088/0031-9120/39/1/006

[9] Luan D, Wang Y, Tang J, Jain D. Frequency distribution in domestic microwave ovens and its influence on heating pattern. Journal of Food Science. 2017;**82**:429-436. DOI: 10.1111/1750-3841.13587

[10] Nissinen TA, Kiros Y, Gasik M, Leskela M. $MnCo_2O_4$ preparation by microwave-assisted route synthesis (MARS) and the effect of carbon admixture. Chemistry of Materials. 2003;**15**:4974-4979

[11] Kubel E Jr. Industrial Heating. Special Focus: Advancements in Microwave Heating Technology [Internet]. 2005. Available from: https://www.industrialheating.com/articles/87800-special-focus-advancements-in-microwave-heating-technology [Accessed: 14 March 2021]

[12] Hussein MZ, Zainal Z, Ming CY. Microwave-assisted synthesis of Zn-Al-layered double hydroxide-sodium dodecyl sulfate nanocomposite. Journal of Materials Science Letters. 2000;**19**: 879-883

[13] Kay E, Gibbons KE, Jones M, Blundell SJ, Mihut A, Gameson I, et al. Rapid synthesis of colossal magnetoresistance manganites by microwave dielectric heating. Chemical Communications. 2000;**2**:159-160

[14] WiFi Wikipedia [Internet]. 2021. Available from: https://en.wikipedia.org/wiki/Wi-Fi [Accessed: 14 March 2021]

[15] Simon CJ, Dupuy DE, Mayo-Smith W. Microwave ablation: Principles and applications. Radiographics. 2005;**25**: S69-S83. DOI: 10.1148/rg.25si055501

[16] Chiu H-M, Sanagavarapu AM, Weily AR, Guy D, Ross D. Analysis of a novel expanded tip wire (ETW)

antenna for microwave ablation of cardiac arrhythmias. IEEE Transactions on Biomedical Engineering. 2003;**50**: 890

[17] Roy R, Peelamedu R, Grimes C, Cheng J, Agrawal D. Major phase transformations and magnetic property changes caused by electromagnetic fields at microwave frequencies. Journal of Materials Research. 2002;**17**:3008-3011

[18] Makul N, Rattanadecho P. Microwave pre-curing of natural rubber-compounding using a rectangular waveguide. International Communications in Heat and Mass Transfer. 2010;**37**:914-923. DOI: 10.1016/j.icheatmasstransfer.2010.03.001

[19] Villegas RV. Non-linear physics of magnetization under microwave and Zeeman excitation in ferromagnetic materials (Física no-lineal de la magnetización bajo excitación Zeeman y de microondas en materiales ferromagnéticos) [doctoral thesis]. Mexico: Instituto Politécnico Nacional; 2019

[20] Microwaving Plastic Waste Can Generate Clean Hydrogen [Internet]. 2021. Available from: https://www.newscientist.com/article/2256822-microwaving-plastic-waste-can-generate-clean-hydrogen/#::text=Chemists%20have%20used%20microwaves%20to,developed%20to%20transform%20the%20plastic [Accessed: 14 March 2021]

[21] Jie X. Microwave initiated catalytic deconstruction of plastic waste into hydrogen. Nature Catalysis. 2020;**3**: 902-912. DOI: 10.1038/s41929-020-00518-5

[22] Booth K. £20M Plastic to Hydrogen Plant at Peel L&Ps Protos Moves Forward, Building-Design & Construction Magazine (BDC). Trafford City, Manchester, UK: Peel (L&P) UK; 2020

[23] Uekert T, Kuehnel M, Wakerley DW, Reisner E. Plastic waste as a feedstock for solar driven H_2 generation. Energy & Environmental Science. 2018;**11**:2853-2857. DOI: 10.1039/C8EE14K408F

[24] Feynman RP, Leighton RB, Sands M. The Feynman Lectures on Physics. Mainly Electromagnetism and Matter. Vol. II. The New Millennium Edition. New York: Basic Books; 2010. p. 566. ISBN: 978-0-465-07998-8

[25] Lorain P, Corson D. Electromagnetic Fields and Waves. 2nd ed. New York: WH Freeman and Company; 1970. p. 696. ISBN: 0-7167-0331-9

[26] Jackson JD. Classical Electrodynamics. New York, London: John Wiley & Sons, Inc.; 1962. p. 656

[27] Griffiths DJ. Introduction to Electrodynamics. 3rd ed. New Jersey, USA: Prentice Hall; 1999. p. 576

[28] Wangsness RK. Electromagnetic Fields. New York, London: John Wiley; 1986. p. 286

[29] Reitz JR, Milford FJ, Christy RW. Foundations of Electromagnetic Theory. 4th ed. Reading, Massachusetts, USA: Adison-Wesley; 2008. p. 345. ISBN: 10-0321581747

[30] Kamol S. Three-dimensional standing waves in a microwave oven. American Journal of Physics. 2010;**78**: 492-495. DOI: 10.1119/1.3329286

[31] Steyn-Ross A. Standing waves in a microwave oven. The Physics Teacher. 1990;**28**:474-476. DOI: 10.1119/1.2343114

[32] Jefimenko OD. Electricity and Magnetism: An Introduction to the Theory of Electric and Magnetic Fields. 2nd ed. New York: Appleton-Century-Crofts; 1966. ISBN: 978-0-917406-08-9

Chapter 7

High-Frequency Electromagnetic Interference Diagnostics

Ling Zhang, Yuru Feng, Jun Fan and Er-Ping Li

Abstract

Electromagnetic interference (EMI) is becoming more troublesome in modern electronic systems due to the continuous increase of communication data rates. This chapter reviews some new methodologies for high-frequency EMI diagnostics in recent researches. Optical modules, as a typical type of gigahertz radiator, are studied in this chapter. First, the dominant radiation modules and EMI coupling paths in an explicit optical module are analyzed using simulation and measurement techniques. Correspondingly, practical mitigation approaches are proposed to suppress the radiation in real product applications. Moreover, an emission source microscopy (ESM) method, which can rapidly localize far-field radiators, is applied to diagnose multiple optical modules and identify the dominant sources. Finally, when numerous optical modules work simultaneously in a large network router, a formula based on statistical analysis can estimate the maximum far-field emission and the probability of passing electromagnetic compatibility (EMC) regulations. This chapter reviews a systematic procedure for EMI diagnostics at high frequencies, including EMI coupling path analysis and mitigation, emission source localization, and radiation estimation using statistical analysis.

Keywords: Electromagnetic interference, Emission source microscopy, Optical modules, High frequency, Industrial products

1. Introduction

Electromagnetic interference (EMI) problems are drawing more and more attention in modern electronic devices and systems. Industrial products have to satisfy electromagnetic compatibility (EMC) requirements, such as US Federal Communications Commission (FCC) regulations. However, the increase in data rate and source complexity is making EMI diagnostics more troublesome. For a sophisticated product with various potential radiation modules, investigating the root cause of radiation, main coupling paths, and practical mitigation approaches is challenging. Moreover, in a complicated system with numerous potential high-frequency radiation sources, identifying the dominant trouble-makers and quantifying the total contribution from these sources is a tedious and time-consuming process. This chapter presents a systematic procedure for high-frequency EMI diagnostics in industrial products by reviewing some recently published methodologies. As a typical high-frequency radiation source, optical modules are studied in this chapter.

First, the interior EMI coupling paths and possible mitigation methods of optical modules are studied [1–6]. The internal mechanism can be analyzed to alleviate the radiation problem at the design stage or propose mitigation solutions. Building full-wave simulation models is a standard method to understand the behind physics. However, it is challenging to run reliable full-wave simulations for intricate structures. The adopted strategy in [1, 2] is to correlate the total radiated power (TRP) in simulation and measurement to ensure a reliable simulation model. Based on the simulation model, the EMI coupling paths are also investigated. Besides, a concept of energy parcels and their trajectories can provide a more intuitive visualization of the coupling paths. Eventually, according to the analysis result of the EMI coupling paths, mitigation solutions are proposed in both simulation and measurement to suppress the EMI coupling paths, which can effectively reduce the far-field radiation to meet EMC regulations.

For a large system with multiple radiators, identifying the dominant sources is a vital but tedious process. Near-field scanning is a standard method for EMI diagnostics [7, 8]. However, detecting near field also includes evanescent waves that do not contribute to far-field radiation. Under some circumstances, near-field probes cannot reach close enough to all locations and components due to mechanical limitations. Emission source microscopy (ESM) is a scanning technique to localize sources contributing to far-field radiation [9]. Sparse ESM [10] is an improved ESM method that identifies the dominant sources through sparse scanning samples. Therefore, sparse ESM is a more efficient method than near-field scanning to determine the main contributor for far-field radiation at high frequencies. In [10], the sparse ESM technique is used to localize the dominant sources among multiple optical modules rapidly. Also, in [11], an absorbing material is utilized to mitigate radiation from the identified sources.

Another problem with large amounts of radiators in a complex system is the required hardware and testing time for EMC regulation tests. Hence, the authors in [12, 13] derived a mathematical relation using statistical analysis to fast predict the maximum radiation from a large system with numerous similar sub-systems, with a level of certainty provided. Also, measurements were performed to validate the estimation.

This chapter provides an insightful review of EMI diagnostic approaches in different aspects at high frequencies, including source modeling, coupling path analysis and visualization, EMI mitigation, source localization, and emission level prediction. The remaining sections of this chapter are organized as follows. In Section 2, the EMI coupling path analysis and mitigation approaches for optical modules are elaborated. The emission source microscopy (ESM) algorithm is then introduced and applied to optical modules in Section 3. In Section 4, the method for estimating the emission level from multiple radiators is explained. Finally, a conclusion is provided.

2. EMI coupling paths analysis and mitigation

2.1 Modeling optical sub-assembly (OSA) module

The diagram of a typical optical link is shown in **Figure 1** [1]. The cage connector and the transceiver module are enclosed by the connector shield cage located on the line card PCB. The cage connector, through which the signal is transmitted, connects the line card PCB and the transceiver circuit board. The signal sent to the transceiver circuit board is further transmitted to the silicon photonic subassembly. The electrical-to-optical (and vice versa) interface assembly converts the electrical signal into an optical signal, which is subsequently transferred to other devices through an optical fiber cable. The cage connector, the optical transceiver module,

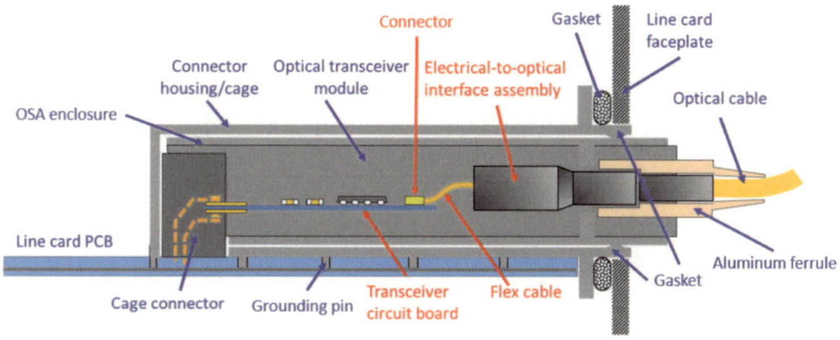

Figure 1.
Structure diagram of the optical transceiver module [1].

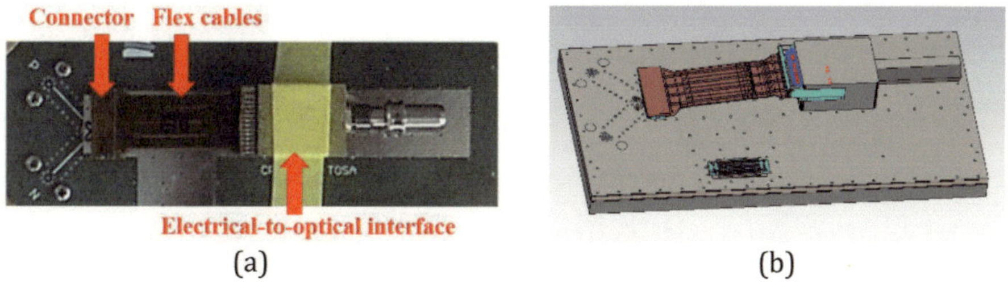

Figure 2.
Optical sub-assembly test board. (a) Experimental test setup. (b) the corresponding simulation model [1].

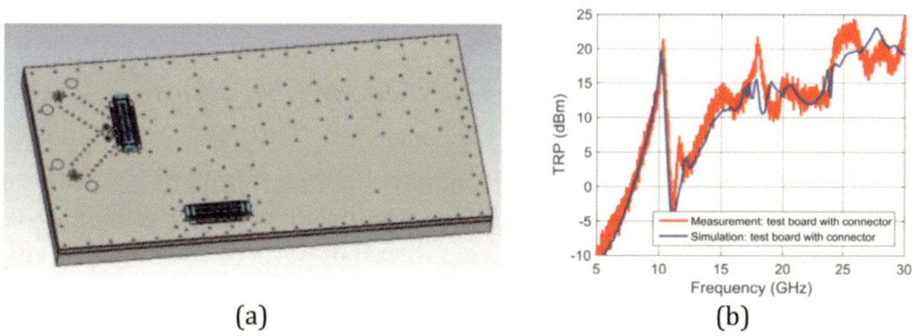

Figure 3.
(a) The simulation model of the test board with the connector only. (b) TRP comparison between measurement and simulation for the test board with the connector only [2].

and the optical cable ferrule were identified as the primary radiation contributors [3, 4]. The aluminum ferrule surrounding the optical fiber cable forms a radiating antenna that is causing the most EMI problems in meeting regulatory requirements for optical transceiver modules.

The optical sub-assembly (OSA) module was modeled in full-wave simulations [2]. The real structure and simulation model for the OSA module is shown in **Figure 2**. There are mainly three parts in the OSA module to be investigated: the connector, the flex cables, and the electrical-to-optical interface assembly. Firstly, the simulation model for the OSA module needs to correlate well with measurements. The adopted strategy [2] was to gradually increase the model complexity and verify the model accuracy step by step. TRP comparison was used for model validation. **Figures 3–5** show the TRP comparison between simulation and measurement for different models. In **Figure 3**, only the connector was kept. In **Figure 4**, the flex cables were included in the model. The entire OSA module was considered in **Figure 5**.

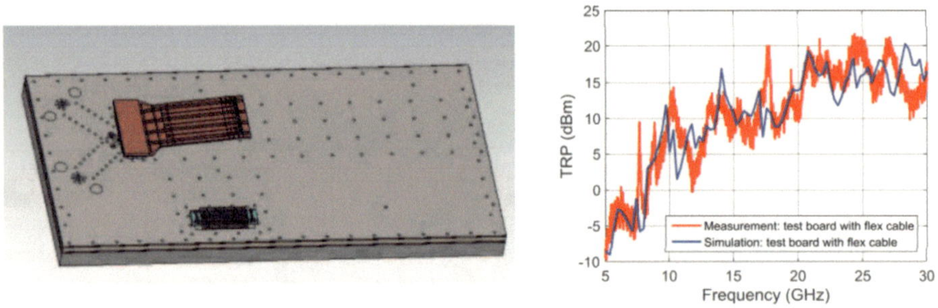

(a) (b)

Figure 4.
(a) The simulation model of the test board with the long flex PCB. (b) TRP comparison between measurement and simulation for the test board with the long flex PCB [2].

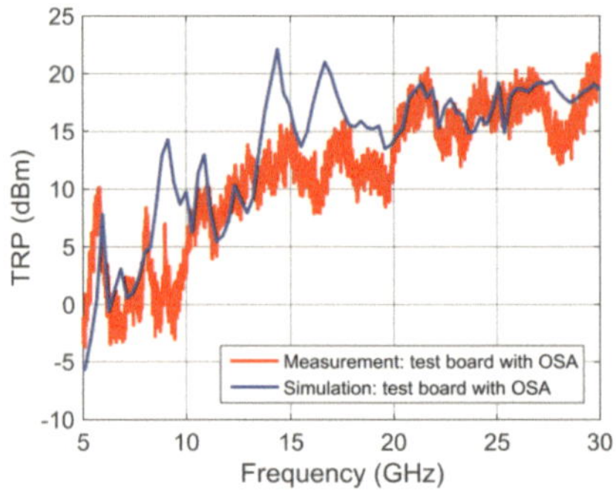

Figure 5.
TRP comparison between measurement and simulation for the test board with OSA model as shown in ***Figure** 2 [2].*

In **Figures 3–5**, the TRP results of simulation and measurement agree reasonably well, indicating the accuracy of the simulation model. Several experiments were designed to quantify the contributions from different parts of the model by utilizing an absorbing material [2], as shown in **Figure 6**. This systematic investigation concludes that the connector and the flex cables are two dominant radiation sources above 10 GHz. In comparison, the electrical-to-optical interface assembly has little contribution to the radiation.

In the OSA module, the connector and the flex cables were diagnosed as the primary radiation contributors above 10 GHz. The structure of the entire optical transceiver module is shown in **Figure 7**, including the OSA module and an enclosure. There are two individual modules in each optical transceiver module – a transmitter optical subassembly (TOSA) and a receiver optical sub-assembly (ROSA) module. Only the TOSA model will be used for analysis since the interior structure of the ROSA cannot be obtained due to confidentiality. A simulation model was built to place the OSA module within an enclosure to mimic the actual design and analyze the coupling paths from the OSA module to the external EMI radiation. This type of optical transceiver module has two working modes at 25.78 GHz and 27.95 GHz. The following EMI analysis will concentrate on 25.78 GHz. The mechanism at 27.95 GHz is similar to 25.78 GHz and hence is omitted.

2.2 Coupling paths analysis by simulations

Figure 8(a) shows a simulation model in which an OSA module is inserted into a shielding enclosure without the enclosure top. One signal trace was excited with a

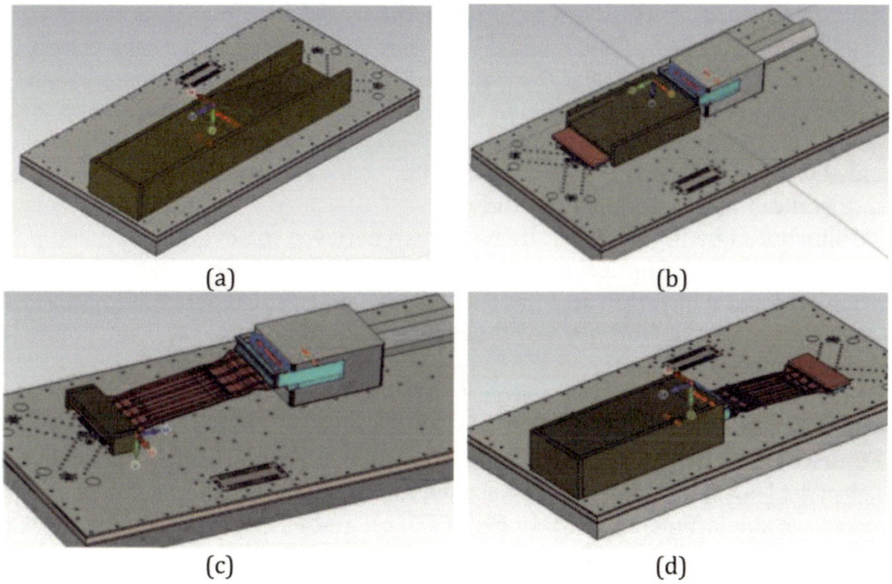

Figure 6.
Use absorbing material to cover different parts of the OSA module to determine their contribution to the radiation. (a) the entire OSA module. (b) the flex cables. (c) the connector. (d) the electrical-to-optical interface assembly [2].

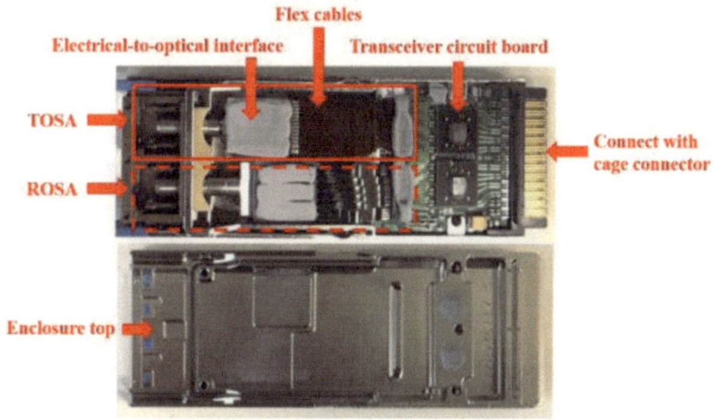

Figure 7.
Structure diagram of the optical transceiver module [1].

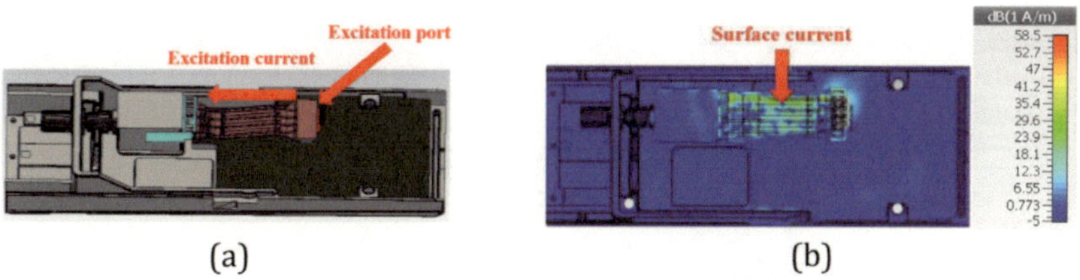

Figure 8.
(a) OSA simulation model only without the enclosure top. (b) Surface current distribution at 25.78 GHz in top view [1].

lumped port. The surface current under this circumstance is plotted in **Figure 8(b)**. The surface current is mainly distributed over the connector and the flex cables, as expected. There is little current on the electrical-to-optical interface assembly. It does not contribute to the radiation of the OSA module, which has been validated by applying absorbing materials.

Subsequently, an enclosure was added to the model in **Figure 8**, as shown in **Figure 9**. The right end of the module in **Figure 7** is inserted into a cage connector on a line card. The leakage near the cage connector is negligible compared to the leakage from the optical fiber cable in the front. Therefore, in the simulation model **Figure 9(a)**, the enclosure end that should be connected with a line card was shielded with PEC. In this manner, the coupling path and radiation from the optical output end can be focused on and studied.

As illustrated by **Figure 9(b)**, there is a gap between the cylindrical metal port and the enclosure. The optical fiber cable is connected to this port. Moreover, a mechanical handle insertion area has a small gap and can also potentially cause EMI leakage. Experimental work proved these gaps to be the primary leakage points, as will be introduced later.

After the enclosure top is added to the simulation model, the surface current distributions are plotted in **Figure 9(c)–(e)**. Compared to the current distribution in **Figure 8(b)**, more current appears in the other places, including the enclosure, the electrical-to-optical interface assembly, and the cylindrical metal port that egresses the shielding enclosure. In **Figure 9(e)**, an evident radiation leakage can be observed from the gap. The simulation comparison between **Figure 9** and **Figure 8** demonstrates that "the enclosure cavity provides an EMI coupling path for propagating modes that can illuminate slots and seams in the OSA enclosure" [1]. The aluminum ferrule of the optical fiber cable inserted into the TOSA port forms a monopole antenna that can be efficiently excited with the current leakage. Typically, a network equipment system has tens of or even hundreds of such monopole antennas, which can cause severe EMI issues in meeting compliance requirements.

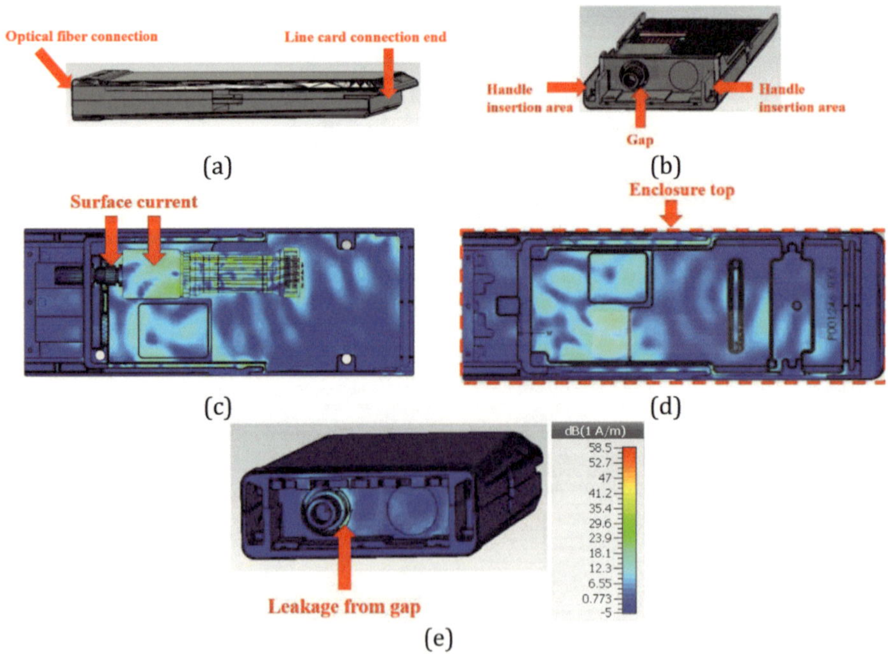

Figure 9.
(a) A simulation model for the optical transceiver module, enclosed in a complete enclosure. (b) the positions of the handle insertion area and the gap between the cylindrical metal part and the enclosure. (c) Surface current distribution on the OSA module and the enclosure bottom at 25.78 GHz, with the enclosure top being hidden. (d) Surface current distribution on the OSA module and the enclosure top at 25.78 GHz, with the enclosure bottom being hidden. (e) Surface current distribution on the enclosure surface in front view at 25.78 GHz [1].

2.3 EMI mitigation of optical transceiver modules

The primary radiation sources and the dominant coupling paths have been determined by using simulation and measurement techniques. In this section, mitigation approaches will be proposed correspondingly to suppress the radiation leakage in both simulation and measurement. The effectiveness of the mitigation methods will, in turn, validate the analyzed coupling paths.

2.3.1 EMI mitigation in simulation

Figure 10 shows simulation models with absorbing material being placed at different locations. **Table 1** compares the simulated TRP for the models in **Figure 10** at 25.78 GHz and 27.95 GHz. After adding absorbing material above the flex cables on the underside of the enclosure lid, a TRP reduction of 4–7 dB is achieved at the two frequencies. The TRP is further reduced by 2–3 dB by adding absorbing material to the cylindrical egress of the electrical-to-optical module and the handle insertion areas. Adding absorbing material on the metal surface of the electrical-to-optical module and the cylinder can significantly reduce the TRP by 8–9 dB.

2.3.2 EMI mitigation in measurement

One chassis with functioning optical transceiver modules did not meet the FCC Class A limit plus a required margin at 25.78 GHz and 27.95 GHz in an EMC test. For example, one measurement showed that one line card with four optical transceiver modules passed the FCC Class A limit by 0.4 dB with no margin typically required in EMC regulations. Based on the analysis result of the dominant coupling paths, some mitigation approaches were implemented in the measurement to suppress the EMI radiation and meet the FCC Class A limit with more margins.

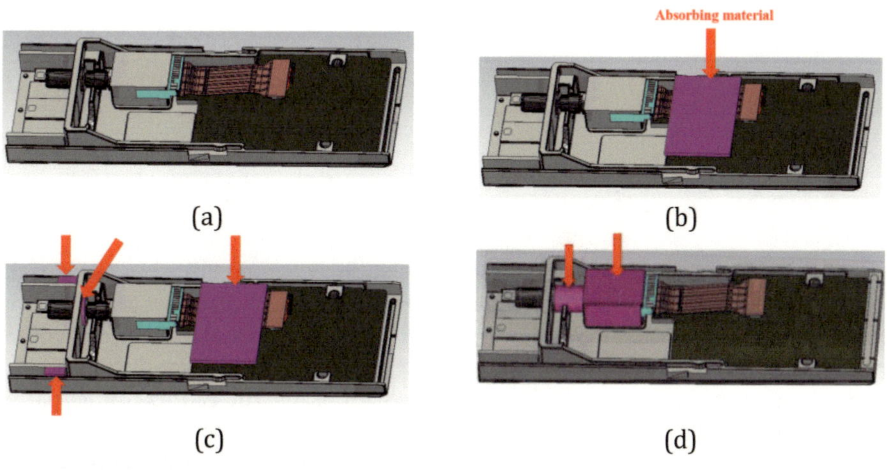

Figure 10.
(a) A simulation model with closed enclosure (enclosure top being hidden) and without absorber. (b) a simulation model with a sealed enclosure, an absorbing material of 26 mm × 16 mm, and a thickness of 1 mm on the enclosure lid underside above the flex cables. (c) a simulation model with a closed enclosure, with absorbing material of 1 mm thickness on the enclosure top above the flex cables, and to the space between the cylindrical metal part of the electrical-to-optical module and the enclosure, as well as the leakage space on the two handle sides of the enclosure. (d) a simulation model with a closed enclosure and absorbing material with a thickness of 0.5 mm on the metal of the electrical-to-optical interface assembly and the metal cylinder inside the enclosure [1].

Simulated TRP (dBm)	25.78 GHz	27.95 GHz
Model in **Figure 10(a)**	−14.9 dBm	−17.3 dBm
Model in **Figure 10(b)** (Compared to **Figure 10(a)**)	−22.5 dBm (7.6 dB reduction)	−21.3 dBm (4 dB reduction)
Model in **Figure 10(c)** (Compared to **Figure 10(a)**)	−25.3 dBm (10.4 dB reduction)	−23.0 dBm (5.7 dB reduction)
Model in **Figure 10(d)** (Compared to **Figure 10(a)**)	−24.0 dBm (9.1 dB reduction)	−25.3 dBm (8 dB reduction)

Table 1.
TRP simulation results of ***Figure 10*** *[1].*

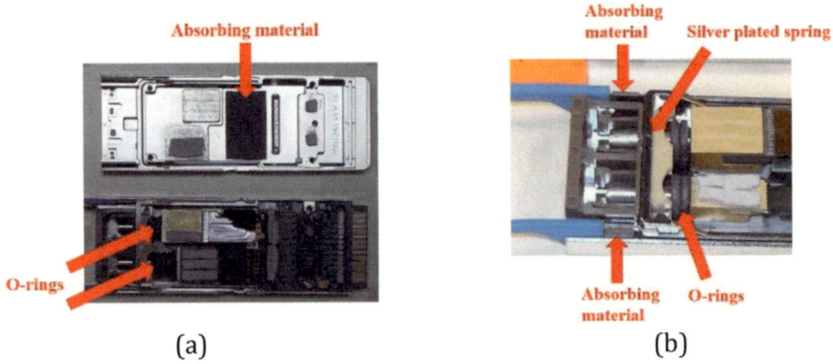

Figure 11.
EMI mitigation methods in the production hardware. (a) Adding absorbing material inside the enclosure above the flex cables and adding grounding O-rings around the cylindrical metal portion of the electrical-to-optical module. (b) Adding some silver-plated spring around the cylindrical metal and adding absorbing material on the handles on two sides [1].

Difference with FCC Class A limit (dB)	25.78 GHz	27.95 GHz
Adding absorbing materials + O-rings (2 line cards, 8 modules)	−8 dB	−4 dB
Adding absorbing materials + O-rings + silver plated spring seal + absorbing materials on handles (2 line cards, 8 modules)	−15 dB	−9 dB

Table 2.
Mitigation of radiated emissions [1].

Figure 11 shows some mitigation methods, such as adding absorbing material, O-rings, and silver-plated spring. **Table 2** shows the EMC test results after applying the mitigation solutions on two line cards with eight optical transceiver modules. Compared with the FCC Class A limit, the largest margin is increased to 15 dB after implementing the mitigations. This outcome corroborates the EMI coupling paths and the mitigation approaches discussed earlier.

A systematic approach has been elaborated to investigate the EMI coupling paths in an optical transceiver module. The procedure includes building simulation models, identifying the dominant radiation modules, analyzing EMI coupling paths through simulations, and finally validating the coupling paths by mitigation in measurements.

The EMI coupling paths in the optical transceiver module can be concluded as follows. The cage connector [4], the connector between the transceiver circuit board and the flex cables, and the flex cables are the dominant radiation sources [1].

DOI: http://dx.doi.org/10.5772/intechopen.97613

The module enclosure provides a cavity-like structure for the propagating modes to radiate from the gap between the electrical-to-optical module and the enclosure. The aluminum ferrule surrounding the optical fiber can be excited with the leakage current and radiate efficiently like a monopole antenna. **Figure 12** illustrates the concluded coupling paths. For a large equipment rack, the EMI can violate regulatory limits with many tens or hundreds of such modules. Mitigation approaches such as absorbing material and elastomer O-rings can effectively suppress EMI radiation in the actual product applications.

2.4 EMI coupling paths visualization

2.4.1 Definition of EM energy parcels

The simulation method in [1] demonstrated the coupling paths, but it did not directly show the coupling paths. Li *et al.* [14, 15] proposed a concept to intuitively visualize coupling paths by utilizing the analogy between the flow of EM energy and fluid flow. The flow of EM energy can be imagined as a certain amount of energy parcels propagating in space. The EM energy flow is defined by the Poynting vector, meaning that the EM energy flow follows the law of conservation of energy. Eq. (1) defines the instantaneous velocity of EM energy parcels.

$$\vec{v}\left({}^{m}\!/_{s}\right) = \frac{\vec{S}(J/m^2s)}{u(J/m^3)}, \tag{1}$$

where Vector S represents the instantaneous Poynting vector and is given by Eq. (2), and u means the total magnetic and electric energy density defined by Eq. (3).

$$\vec{S} = \vec{E} \times \vec{H}. \tag{2}$$

$$u = \mu\frac{\vec{H}\cdot\vec{H}}{2} + \varepsilon\frac{\vec{E}\cdot\vec{E}}{2}. \tag{3}$$

The time-averaged velocity of the energy parcels is defined as Eq. (4), which is a constant value.

$$\vec{v}_{av} = \frac{\overrightarrow{S_{av}}}{u} = \frac{\mathrm{Re}\left[\vec{S}\right]}{u}. \tag{4}$$

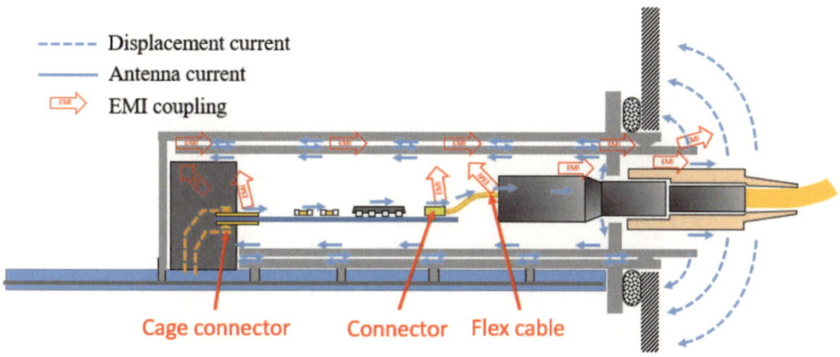

Figure 12.
Summary of the EMI coupling paths in the optical transceiver module [1].

Therefore, the trajectory of the energy parcels can be acquired by calculating the streamline (tangential line) of the real part of the complex Poynting vector [14, 15]. The trajectories are tracked back from the receiver antenna to visualize the energy path from the transmitter to the receiver.

2.4.2 Apply coupling path visualization to optical modules

Some researchers applied the trajectory concept of energy parcels to visualize the EMI coupling paths on quad form-factor pluggable (QSFP) modules and propose corresponding mitigation approaches [5, 6]. In [5], an EMI issue related to a QSFP module possessing a heatsink was studied. The issue is that when an optical module is inserted, the heatsink on top of the optical module rises. Thus an air gap is created by the rising heatsink, and the shielding effectiveness (SE) of the QSFP shielding cage is degraded. **Figure 13** describes the measurement setup to quantify the SE of the QSFP shielding cage. There were two reverberation chambers (RC): one side is noisy, and the other side is quiet. A transmitter (Tx) antenna and a receiver (Rx) antenna were placed on the noisy and quiet sides, respectively.

The potential coupling paths related to the rising heatsink are depicted in **Figure 14**. The rising heatsink provides a guiding structure for the propagating waves. The EM waves can penetrate the cage from the path between the heatsink and the cage. This path around the cage guides the EM waves through the cage. Besides, there is a small gap between the cage and the line card PCB so that EM waves can go through the cage from the bottom.

The concept of energy parcels and their trajectories was applied to validate the above hypothesis about the coupling paths [5]. The result of the trajectories is shown in **Figure 15**. It can be observed that the rising heatsink behaves as a guided

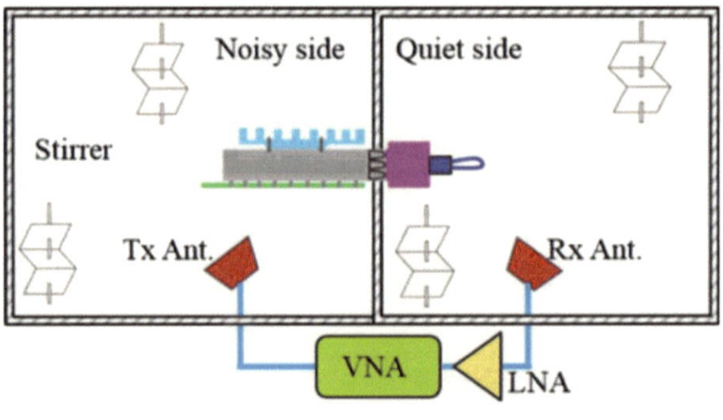

Figure 13.
Measurement setup for the shielding effectiveness of the optical module [5].

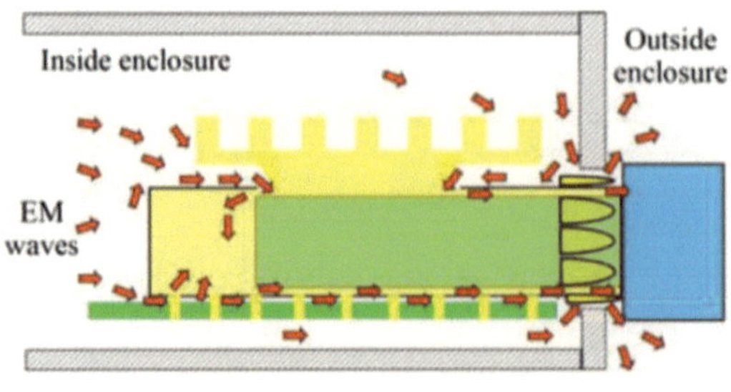

Figure 14.
Potential coupling paths when there is a heatsink [5].

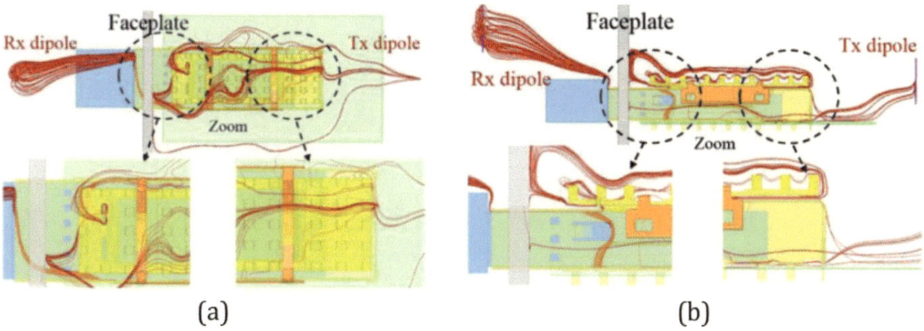

Figure 15.
Reversed tracked energy parcels from the receiver antenna (Rx) to the transmitter antenna (Tx). (a) Top view. (b) Side view [5].

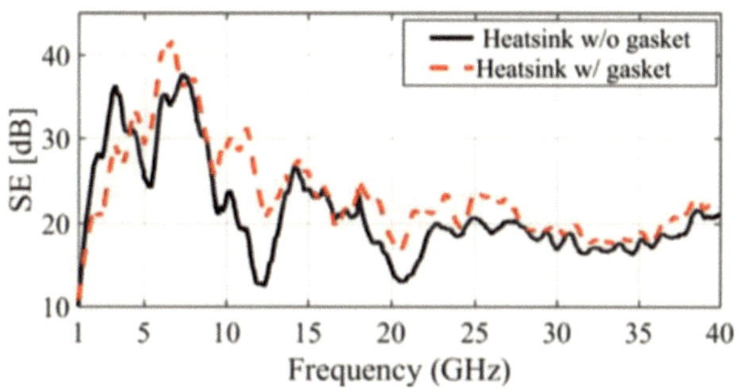

Figure 16.
Comparison between the averaged SE of the cage with and without the gasket [5].

structure for the propagating waves, and the area between the heatsink and the top of the cage contributes as a leakage path.

According to the analysis result using the coupling path visualization, a mitigation approach was applied by placing a window frame gasket between the cage and the heatsink [5]. The change of SE is plotted in **Figure 16**. The SE of the QSFP shielding cage is notably improved by a few decibels over the entire frequency range. This improvement corroborates the EMI coupling paths concluded from the trajectory visualization of energy parcels. Similarly, the same approach was adopted in [6] to investigate the coupling paths for flyover QSFP connectors. Also, corresponding mitigation methods using an absorbing material were demonstrated to work effectively.

3. Emission source microscopy (ESM)

The emission source microscopy (ESM) method [9] can identify far-field radiation sources by scanning the far field over a plane above a device under test (DUT) and back-propagating the field onto the DUT plane to localize the dominating sources. Both the field magnitude and phase are required for the back-propagation calculation. Therefore, phase measurement is needed. A typical way of measuring the field phase from an active DUT is shown in **Figure 17**. Antenna B is the reference antenna with a fixed location for phase measurement. Antenna A is the scanning antenna that moves with a scanning robot to acquire the field information at different locations. Antenna A and B are connected to the channel A and B of a vector network analyzer (VNA). The VNA operates at the tuned receiver mode and

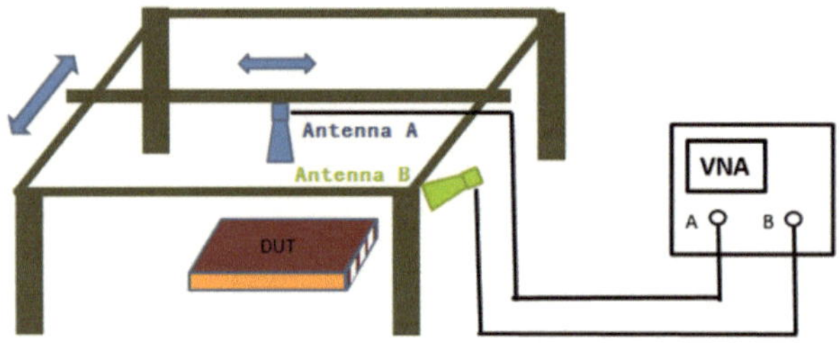

Figure 17.
ESM system setup [10].

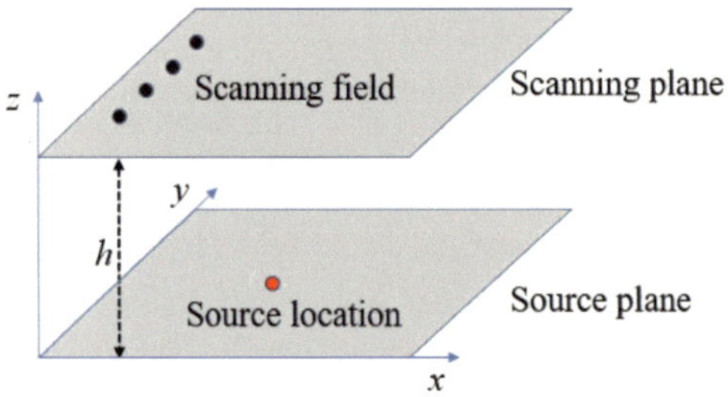

Figure 18.
Diagram of ESM algorithm.

receives signals at channels A and B simultaneously. The phase difference between A and B is used as the field phase, while the signal magnitude received by channel A is the field magnitude.

3.1 Algorithm description

Figure 18 illustrates the ESM algorithm. There is a scanning plane and a source plane, which are parallel to the x-y plane. The distance between them is h. The far field, including magnitude and phase, is collected on the scanning plane. Suppose that $E_t(x, y, h)$ is the tangential field on the scanning plane, and $E_t(x, y, 0)$ is the tangential field on the source plane. The mathematical calculation of the ESM algorithm [9] can be summarized by Eq. (5).

$$E_t(x, y, 0) = F^{-1}\{F[E_t(x, y, h)] \cdot e^{jk_z h}\}, \tag{5}$$

where $k_z = \sqrt{k^2 - k_x^2 - k_y^2}$ is the z component of the propagation vector if $k_x^2 + k_y^2 \leq k^2$; $k_z = j\sqrt{k_x^2 + k_y^2 - k^2}$ if $k_x^2 + k_y^2 > k^2$; k_x and k_y are the spatial frequencies of 2D Fourier transform; F and F^{-1} are the forward and reverse 2D Fourier transform operators, respectively. The basic idea behind this ESM method is to decompose the scanning field into plane waves with different propagation vectors in x, y, and z directions.

In [11], an electronic system with multiple physical layer transceivers was diagnosed using the ESM method. The source location at a frequency of interest was precisely localized. Afterward, by applying an absorbing material to the identified source location, the radiated power at the corresponding frequency was effectively suppressed.

3.2 Sparse ESM

Using the original ESM method [9], dense and uniform scanning samples are needed, which is time-consuming and inefficient in fast EMI diagnostic applications. Therefore, a sparse ESM method [10] was proposed to improve the scanning speed using sparse samples. Even though background noise is introduced to the reconstructed images, the scanning time is significantly reduced. Mathematical derivations show that the signal-to-noise ratio (SNR) of the reconstructed images is proportional to the number of sparse samples. Thus, the dominant radiation sources can be identified using an appropriate number of scanning points.

The sparse ESM method can locate dominant sources, but the absolute field strength of the reconstructed images is not trustable. More sparse samples will increase the field magnitude of the reconstructed images due to the mathematical process of Fourier transform. To tackle this limitation, a nearest-neighbor interpolation method [10] was adopted to interpolate the scanning field and calculate the radiated power from the sources. **Figure 19** shows an example of the nearest-neighbor interpolation method. According to the sparse samples, the scanning domain is segmented into multiple polygon areas, and each polygon represents the area closest to one scanning sample. Afterward, the radiated power through the scanning plane can be calculated according to Poynting's theorem expressed in Eq. (6) [10].

$$P_{tot} = \iint_A \frac{\left|\vec{E}\right|^2}{2\eta} \cdot dS = \sum_i \frac{\left|\vec{E}_i\right|^2}{2\eta} \cdot A_i, \tag{6}$$

where η is the free space wave impedance; $\vec{E}_i$ is the field at sampling location i; A_i is the area of the corresponding nearest-neighbor region.

To validate if radiated power can be accurately calculated, an experiment was performed [10] using the setup in **Figure 20**. Two horn antennas were utilized as the transmitter and receiver, respectively. Antenna B was at a fixed position, and Antenna A was moved manually through a scanning robot to obtain the radiated field of Antenna B. Assume that the main beam of Antenna B is narrow, which is true in most cases. The scanning plane is sufficiently large compared with the main beam of Antenna B. Hence, the radiated power of Antenna B can all propagate through the scanning plane. Moreover, the polarization direction of Antenna A and B are aligned with each other. Thus, it is reasonable to expect that the radiated

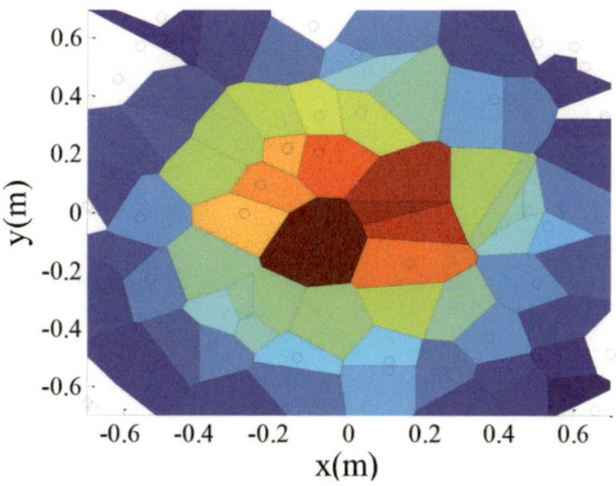

Figure 19.
An example of the nearest-neighbor interpolation [10].

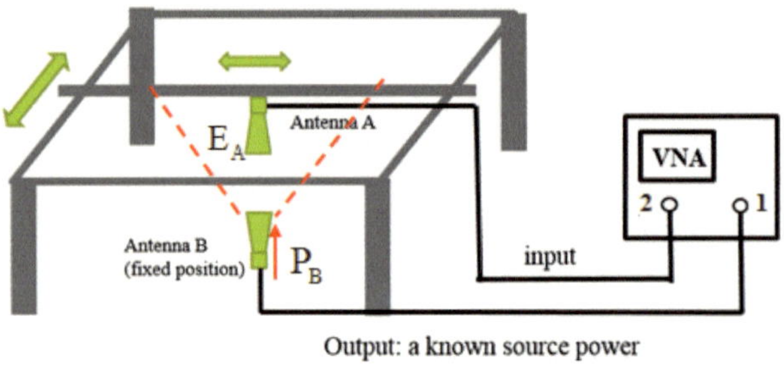

Figure 20.
Measurement setup for validating the nearest-neighbor interpolation method in calculating the radiated power [10].

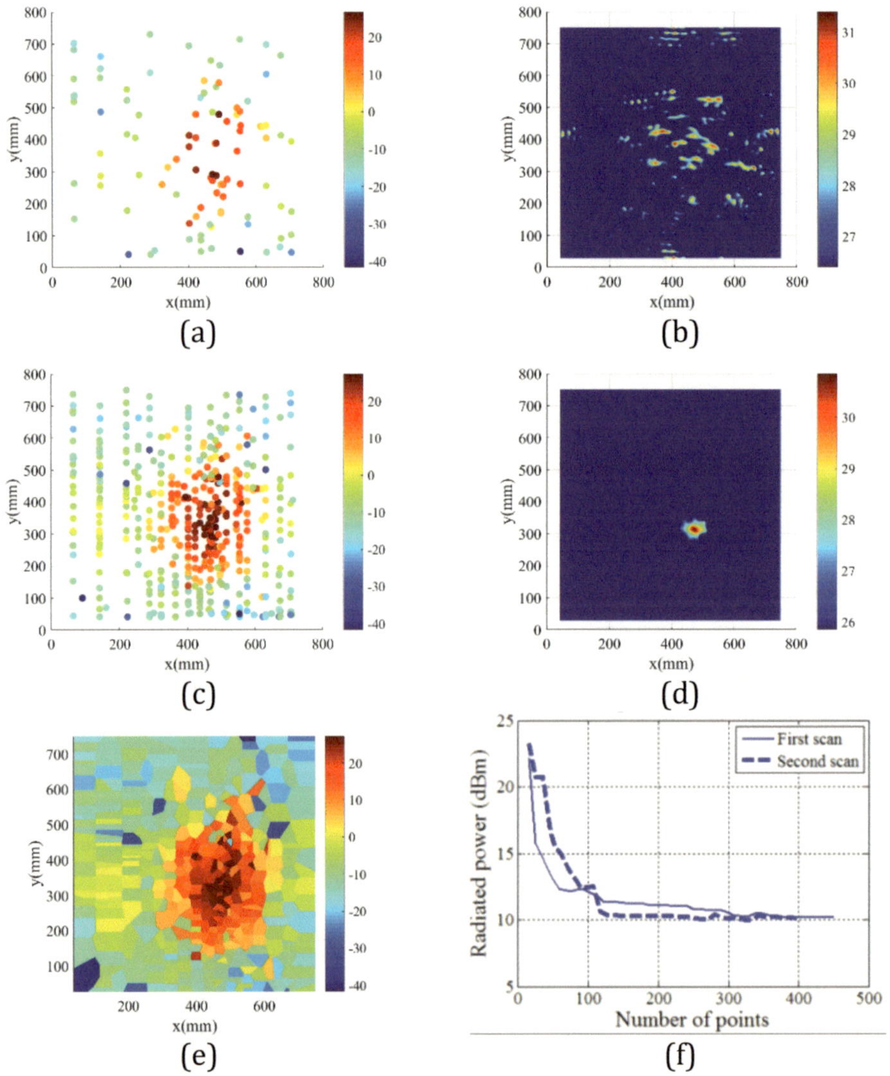

Figure 21.
(a) Scanning samples when N = 100. (b) Reconstructed image when N = 100. (c) Scanning samples when N = 454. (d) Reconstructed image when N = 454. (e) E field magnitude on the scanning plane after using nearest-neighbor interpolation when N = 454. (f) the convergence of calculated radiated power as a function of the number of samples [10].

power of Antenna B can be all captured by Antenna A using the nearest-neighbor interpolation method.

A vector network analyzer (VNA) was utilized to measure the scanning field through S_{21} as shown in **Figure 20**. The measured E field values of Antenna A can be obtained through the antenna factor of Antenna A. If the incident voltage of Antenna B is 1 V, the incident power of Antenna B will be calculated as 10 dBm in a 50-Ω transmission line system. In other words, the expected radiated power calculated using the nearest-neighbor interpolation method is 10 dBm.

Figure 21 presents the measurement process using the measurement setup in **Figure 20**. With the increase of sampling number, the quality of the reconstructed image is also improved according to the conclusion in [10]. **Figure 21(e)** shows the scanning field after using the nearest-neighbor interpolation method. **Figure 21(f)** plots the convergence of the calculated radiated power for two different scans, which both converged to 10 dBm as expected. This experiment demonstrates that the sparse ESM method can locate dominant radiation sources and estimate radiated power by sparse scanning samples.

3.3 Apply ESM to optical transceiver modules

Optical transceiver modules are widely used in Gigabit Ethernet systems to transmit high-frequency data of several hundred gigabits per second (Gbps). Serious EMI problems can be caused at high frequencies by a large number of these modules. EMI diagnostics of multiple potential radiation modules is troublesome. The sparse ESM method introduced earlier is a valuable tool to locate radiation sources efficiently and quantify radiated power. In this section, the sparse ESM method is applied to diagnose multiple optical transceiver modules. **Figure 22** shows a real-world DUT with 16 optical transceiver modules, which generate 15 distinct radiation peaks around 10.3125 GHz. The frequency peak # 11 was selected in this measurement. The sparse ESM was adopted to identify the optical transceiver module that caused frequency # 11 among the 16 modules.

A zero span of a VNA was used, as explained in [9], to focus on frequency # 11. The field magnitude and phase on a scanning plane were measured using the VNA tuned-receiver mode, as illustrated in **Figure 17**. **Figure 23** shows the measured field, reconstructed image, and calculated radiated power. The total number of scanning samples was 352.

In **Figure 23(b)**, the reconstructed source image has a clear hotspot corresponding to the location of one pair of optical transceiver modules. Interestingly, as shown in **Figure 23(c)**, the reconstructed field phase shows a series of phase contours surrounding the corresponding source location. Even though the sampled scanning field was obtained sparsely, the reconstructed image has a

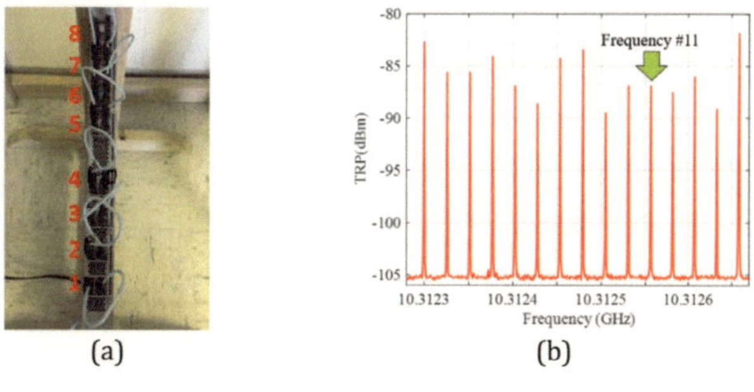

Figure 22.
(a) Validation DUT with 16 optical transceiver modules (numbered by pairs). (b) 15 radiation peaks from the DUT around 10.3125 GHz [10].

continuous distribution due to Fourier transform and inverse Fourier transform in Eq. (1). **Figure 23(d)** shows the convergence of radiated power as more scanning points were sampled. The converged radiated power is approximately −83.4 dBm.

One more experiment was implemented to validate the identified radiation module in **Figure 23** by removing radiation pair # 3 from the DUT and performing another ESM scan. The result is shown in **Figure 24**. The measured scanning field, the reconstructed source field, and the calculated radiated power were significantly reduced. After removing the radiation module for frequency # 11, there is some radiation from the same location. The reason is that the physical layer interface IC beneath the optical module is still radiating.

Furthermore, the total radiated power (TRP) for the two cases in **Figures 23** and **24** was measured in a reverberation chamber (RC) for comparison, and the result is listed in **Table 3**. After removing radiated pair # 3, the TRP value is reduced by 14.1 dB, which is close to the 13.5 dB reduction estimated by the sparse ESM scanning.

The above experiments demonstrate that the sparse ESM is a reliable technique to identify the locations and quantify the radiated power of dominant sources at high frequencies. As a commonly used product in data communication systems, optical transceiver modules were used as the DUT. The sparse ESM method can be applied to identify the radiating modules rapidly.

However, the measurement for the sparse ESM method mentioned above was performed manually, which was user-unfriendly. Also, the image quality can be degraded by operators' subjective decisions and perceptions. An approach based on Gaussian process regression was proposed to achieve automatic scanning to address this issue [16]. The Gaussian process is used to estimate the amplitude variation of the scanned field and its uncertainty to focus on the most relevant scanning area. Thus, compared with pure random scanning, the proposed approach can save the number of scanning samples.

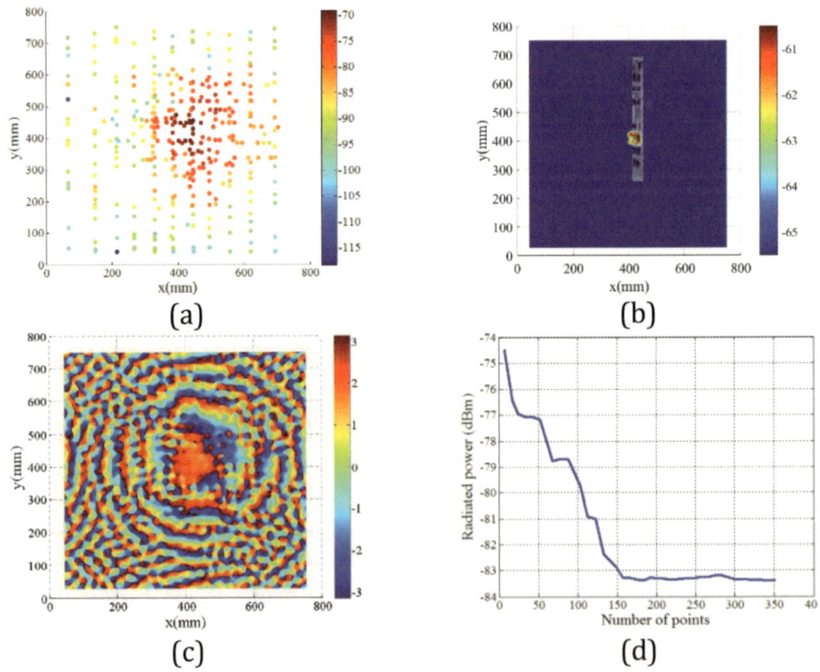

Figure 23.
Scanning result at frequency # 11. (a) E field on the scanning plane. (b) the magnitude of the reconstructed image on the DUT plane. (c) the phase of the reconstructed image on the DUT plane. (d) the convergence of calculated radiated power [10].

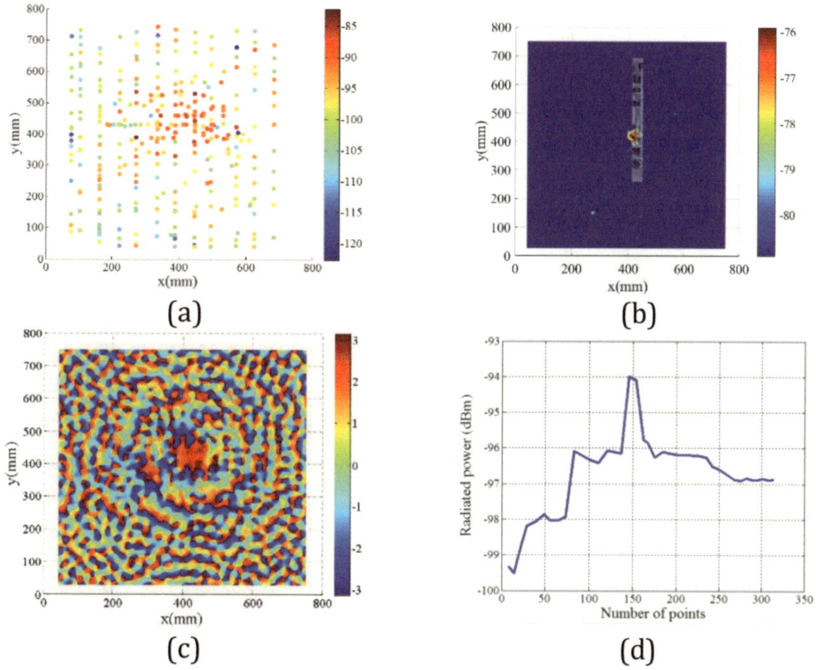

Figure 24.
Scanning result at frequency # 11 after removing the corresponding radiated pair # 3. (a) E field on scanning plane. (b) the magnitude of the reconstructed image on the DUT plane. (c) the phase of the reconstructed image on the DUT plane. (d) the convergence of calculated radiated power with the increase of the number of samples [10].

Power values	Power estimation through interpolation	TRP measured in an RC
Power with pair # 3 (dBm)	−83.4	−86.9
Power without pair # 3 (dBm)	−96.9	−101.0
Power reduction (dB)	13.5	14.1

Table 3.
Power comparison at frequency # 11 [10].

4. Multiple EMI radiators

Measuring the maximum E field emission for a complex system with numerous optical modules requires exhaustive hardware testing. **Figure 25** illustrates a large network router with many optical modules. The 3D radiation pattern of a single optical module was measured as depicted in **Figure 25(b)**. Statistical analysis and simulation were then performed to fast estimate the E field emission from multiple radiators without running actual hardware testing [12, 13].

The phased array antenna theory and Monte Carlo simulation were utilized to perform the statistical analysis [12]. A tendency of $10 \log N$ (dB) was found for the maximum electric field emission from a system with radiators at the same frequency with random phase distribution, where N is the number of radiators. Also, it was found that when multiple radiators with the same magnitude and random distribution contribute simultaneously, the maximum E field follows a normal distribution. Hence, a cumulative distribution function (CDF) can be obtained, and a level of certainty for the E field estimation can be provided.

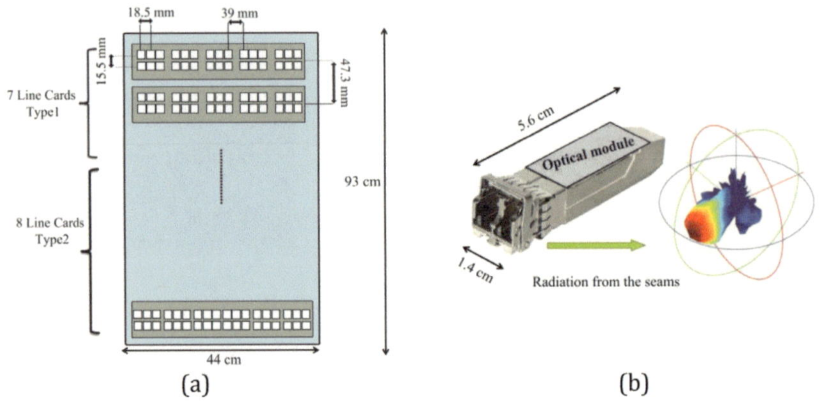

Figure 25.
(a) A large network router consisting of numerous optical modules radiating at 10.31 GHz. (b) a typical optical module and its 3D radiation pattern from the seams [12].

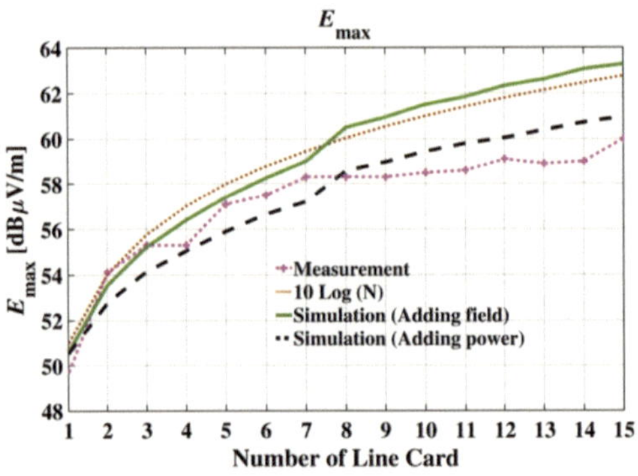

Figure 26.
E_{max} comparison between measurement and prediction as the number of line cards increases [12].

Figure 26 compares the E_{max} between the measurement, simulation, and estimation when the number of line cards increases. "Adding field" means using the same frequency for all line cards, while "adding power" means using different frequencies for each line card. It is shown in **Figure 26** that adding power intensities instead of adding fields has a slower slope than $10logN$ (dB) tendency. From **Figure 26**, it can be concluded that the simulation and measurement results can agree reasonably well with the $10logN$ (dB) tendency. When the number of line cards increases, the measurement result starts to deviate slightly from the estimation. The reason given in [12] is that more line cards meant a higher probability of missing the maximum electric field in the measurement, depending on the radiation pattern of the receiving horn antenna.

More experiments were implemented in [13] to validate the $10logN$ tendency by using patch antennas to mimic the radiation of optical modules. The same agreement between measurement and estimation was observed. **Figure 27** shows the CDF distribution for 15 and 30 patch antennas with random phase excitation. The measured E_{max} under different phase randomizations fell into the 40–90% range. Therefore, it can be concluded from the experiment that the statistical model can reliably estimate the worst E field emission from multiple radiators with random phase distribution. By adopting the $10logN$ (dB) tendency and the CDF of normal distribution, one can easily estimate the likelihood of passing an EMC regulation

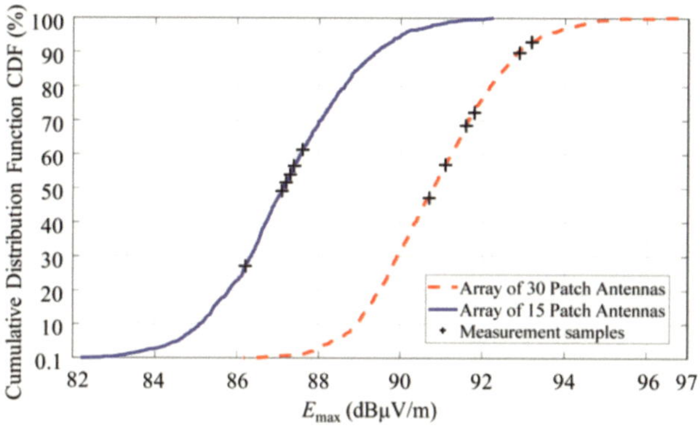

Figure 27.
Measurement samples and prediction of the CDF for two different number of patch antennas [13].

without running exhaustive hardware testing with more radiation modules to be installed.

5. Conclusion

In this chapter, some effective diagnostic approaches for industrial products at high frequencies are reviewed. First, the study of an explicit optical transceiver module is presented to investigate the internal EMI coupling mechanism. An accurate simulation model for the optical subassembly (OSA) module was constructed by correlating the total radiated power of simulation and measurement. More simulations and measurements were performed to identify the dominant radiation modules and coupling paths. An intuitive visualization method for EMI coupling paths was also adopted. According to the EMI coupling paths, mitigation solutions were correspondingly proposed and demonstrated to work effectively in real product applications. Subsequently, an emission source microscopy (ESM) method that can efficiently localize far field radiation sources is introduced. Using the ESM technique, the dominant radiation modules among multiple optical modules can be rapidly identified. Moreover, for numerous sources radiating at the same frequency with random phase distribution, the maximum electric field emission has a $10logN$ (dB) tendency, where N is the number of sources. Thus, the probability of passing an EMC regulation can be fast estimated without running exhaustive hardware testing. This chapter reviews and summarizes an insightful and systematic approach for high-frequency EMI diagnostics for industrial products.

Acknowledgements

The authors would like to thank Prof. James Drewniak, Prof. David Pommerenke, Prof. Victor Khilkevich, Dr. Kyoungchoul Koo, Dr. Jing Li, Dr. Alpesh Bhobe, Xiao Li, Xiangyang Jiao, and Sukhjinder Toor for their kind help and support on the related research.

Conflict of interest

The authors declare no conflict of interest.

Author details

Ling Zhang[1*], Yuru Feng[1], Jun Fan[2] and Er-Ping Li[1]

1 Zhejiang University, Hangzhou, China

2 Missouri University of Science and Technology, USA

*Address all correspondence to: lingzhang_linson@qq.com

References

[1] L. Zhang et al., "EMI Coupling Paths and Mitigation in Optical Transceiver Modules," in IEEE Transactions on Electromagnetic Compatibility, vol. 59, no. 6, pp. 1848–1855, Dec. 2017, DOI: 10.1109/TEMC.2017.2697761.

[2] J. Li et al., "EMI coupling paths in silicon optical sub-assembly package," 2016 IEEE International Symposium on Electromagnetic Compatibility (EMC), Ottawa, ON, Canada, 2016, pp. 890–895, DOI: 10.1109/ISEMC.2016.7571768.

[3] J. Li, S. Toor, A. Bhobe, J. L. Drewniak and J. Fan, "Radiation physics and EMI coupling path determination for optical links," 2014 IEEE International Symposium on Electromagnetic Compatibility (EMC), Raleigh, NC, USA, 2014, pp. 576-581, DOI: 10.1109/ISEMC.2014.6899037.

[4] J. Li, X. Li, S. Toor, H. Fan, A. U. Bhobe, J. Fan, and J. L. Drewniak, "EMI Coupling Paths and Mitigation in a Board-to-Board Connector," in IEEE Transactions on Electromagnetic Compatibility, vol. 57, no. 4, pp. 771–779, Aug. 2015.

[5] A. Talebzadeh, P. C. Sochoux, J. Li, Q. Liu, K. Ghosh, and D. Pommerenke, "Shielding Effectiveness, Coupling Path, and EMI Mitigation for QSFP Cages With Heatsink," in IEEE Transactions on Electromagnetic Compatibility, vol. 60, no. 5, pp. 1254–1262, Oct. 2018, DOI: 10.1109/TEMC.2018.2813889.

[6] A. Talebzadeh et al., "Coupling Path Visualization and EMI Mitigation for Flyover QSFP Connectors," in IEEE Transactions on Electromagnetic Compatibility, vol. 62, no. 4, pp. 1037–1044, Aug. 2020, DOI: 10.1109/TEMC.2019.2943290.

[7] D. Baudry, C. Arcambal, A. Louis, B. Mazari, and P. Eudeline, "Applications of the Near-Field Techniques in EMC Investigations," in IEEE Transactions on Electromagnetic Compatibility, vol. 49, no. 3, pp. 485–493, Aug. 2007, DOI: 10.1109/TEMC.2007.902194.

[8] J. J. H. Wang, "An examination of the theory and practices of planar near-field measurement," in IEEE Transactions on Antennas and Propagation, vol. 36, no. 6, pp. 746–753, June 1988, DOI: 10.1109/8.1176.

[9] P. Maheshwari, H. Kajbaf, V. V. Khilkevich, and D. Pommerenke, "Emission Source Microscopy Technique for EMI Source Localization," in IEEE Transactions on Electromagnetic Compatibility, vol. 58, no. 3, pp. 729–737, June 2016, DOI: 10.1109/TEMC.2016.2524594.

[10] L. Zhang et al., "Sparse Emission Source Microscopy for Rapid Emission Source Imaging," in IEEE Transactions on Electromagnetic Compatibility, vol. 59, no. 2, pp. 729–738, April 2017. DOI: 10.1109/TEMC.2016.2639526.

[11] X. Jiao et al., "EMI mitigation with lossy material at 10 GHz," 2014 IEEE International Symposium on Electromagnetic Compatibility (EMC), Raleigh, NC, USA, 2014, pp. 150–154, DOI: 10.1109/ISEMC.2014.6898960.

[12] J. Meiguni et al., "EMI Prediction of Multiple Radiators," in IEEE Transactions on Electromagnetic Compatibility, vol. 62, no. 2, pp. 415–424, April 2020, DOI: 10.1109/TEMC.2019.2914240.

[13] W. Zhang et al., "System-Level EMI of an Artificial Router System With Multiple Radiators: Prediction and Validation," in IEEE Transactions on Electromagnetic Compatibility, vol. 62, no. 4, pp. 1601–1610, Aug. 2020, DOI: 10.1109/TEMC.2020.3006517.

[14] H. Li, V. V. Khilkevich and D. Pommerenke, "Identification and

Visualization of Coupling Paths—Part I: Energy Parcel and Its Trajectory," in IEEE Transactions on Electromagnetic Compatibility, vol. 56, no. 3, pp. 622–629, 1 June 2014, DOI: 10.1109/TEMC.2014.2314645.

[15] H. Li, V. V. Khilkevich and D. Pommerenke, "Identification and Visualization of Coupling Paths--Part II: Practical Application," in IEEE Transactions on Electromagnetic Compatibility, vol. 56, no. 3, pp. 630–637, 1 June 2014, DOI: 10.1109/TEMC.2014.2314660.

[16] J. Li, J. Zhou, S. Yong, Y. Liu, and V. Khilkevich, "Automatic sparse ESM scan using Gaussian process regression," 2020 IEEE International Symposium on Electromagnetic Compatibility & Signal/Power Integrity (EMCSI), Reno, NV, USA, 2020, pp. 671–675, DOI: 10.1109/EMCSI38923.2020.9191463.

Chapter 8

UHF RFID in a Metallic Harsh Environment

Renata Rampim and Ivan de Pieri Baladei

Abstract

The use of the UHF RFID system in a warehouse that contains steel coils is a challenge for the technology itself since there are countless reflections of the radio frequency waves in the environment causing the multipath effect, which represents one of the most complex problems for wireless communication. Thus, this effect must be managed with the hardware, such as the antenna radiation diagram, and with the software, in the middleware, the software developed for the application. In this chapter, an application of RFID in metallic items will be discussed, tracking them since the product's hot rolling, controlling the receiving process, shipping, and inventory, working together with equipment such as overhead crane.

Keywords: RFID, UHF RFID system, logistic, harsh environment, RFID in metallic

1. Introduction

RFID (Radio Frequency Identification) is an identification technology that uses a magnetic field or electromagnetic waves to access data stored in a microchip that is connected to a small antenna attached to an object.

The RFID system uses the magnetic field at LF (125 kHz) or HF (13.56 MHz) frequencies for communication between the transmitter and receiver. This communication typically has a maximum range of 20 cm, named NFC (Near Field Communication), that is, it is a short-range wireless connectivity technology, which transfers energy to the tag through inductive coupling.

However, for the frequencies used in the RFID system in the UHF band (860 MHz - 960 MHz) there is also a magnetic field constituted near the transmission antenna (near field), which is well defined at this distance. As the magnetic field spreads, an electric field develops. These fields, magnetic (near field) and electric (developed), add up orthogonally and the result is an electromagnetic field. This electromagnetic field has the property of propagation and propagates in the form of an electromagnetic wave, moving away from the transmitting antenna. In this way, when a UHF RFID tag is hit by this electromagnetic wave, coupling occurs. In other words, the energy of the electromagnetic wave captured by the RFID tag antenna energizes the microchip (IC - Integrated Circuit). With this, the IC internally performs its functions and returns its data, object identification and other information that are stored in its memory, to the RFID UHF reader through another electromagnetic wave created by the IC. This description refers to a backscatter coupling.

Typically, due to physical properties, the RFID system at LF and HF frequencies have an easier time reading their tags when they are on objects containing metal, liquids, wood and due to their type of coupling (inductive coupling), which does

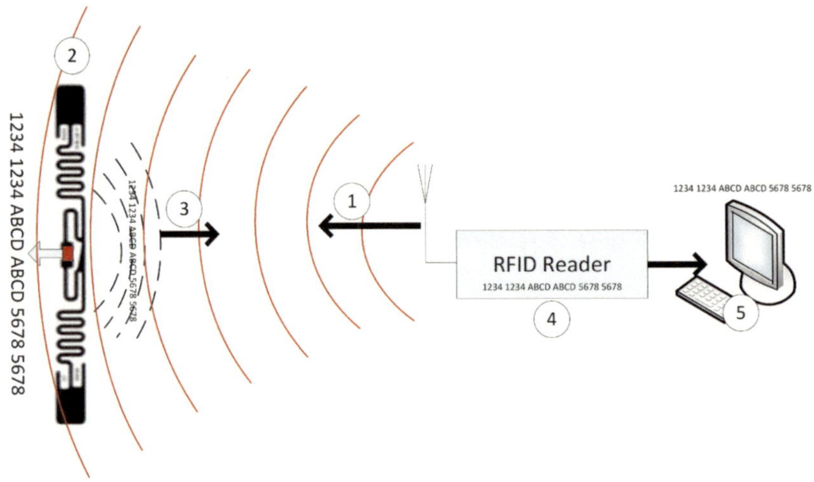

Figure 1.
Overview of an RFID system with backscatter coupling.

not occur with the UHF RFID system. The electromagnetic wave that propagates is strongly influenced by the environment in which the RFID system is implemented and is dominated by the same phenomena that characterize any radio signal, that is, reflection, diffraction, and refraction. These phenomena can alter the amplitude, phase or frequency characteristics of the electromagnetic wave and cause signals of multiple paths, making it difficult to implement the UHF RFID system.

The great challenge of implementing a UHF RFID system in a metallic environment is to control these multipath signals, in which the variables in the environment are numerous and which cause a very unstable signal level, both from the reader to the tag and from the tag to the reader.

1.1 Overview of a UHF RFID system

The overview of an RFID system with backscatter coupling is shown in **Figure 1** [1]. In this figure, at the beginning of the process (1), the RFID reader creates an electromagnetic wave and transmits it.

When the electromagnetic wave meets an RFID tag, the backscatter coupling occurs (2). With the energy resulting from the coupling, the microchip (IC) performs its functions and sends the contents of its memory to the reader through another wave created by it (3).

Upon receiving the wave with the IC memory, the reader demodulates the signal and obtains the data and amplifies them to send to the computer (4).

With this data, the computer performs its functions, such as grouping, filtering and, in this way, prepares them for the application (5).

2. UHF RFID system in a metallic environment

A typical RFID system is divided into two layers: physical and software.

In the physical layer, there are readers, also called interrogators, which functions are: to generate radio frequency energy and query signals and send them through one or more antennas; receiving replies to the queries from RFID tags, amplifying and demodulating these signals; organizing the data received from the RFID tags and, finally, sending them to a computer or a network.

In this way, a reader must have at least one antenna, which is the communication interface with an RFID tag, and an interface for communication with the computer, which can be a serial or USB output, an Ethernet output, Wi-Fi, Bluetooth, 3G.

RFID tags also belong to the physical layer and consist of an inlay, which, in turn, consists of: a microchip or integrated circuit (IC). The data, the unique identification of the object and other information, are stored in the memory of this IC, which is also responsible for several essential processes for communication with the reader to occur; antenna: responsible for receiving and sending radio frequency waves. The shape of the antenna depends on the frequency of operation of the system; connectors: connect the IC to the antenna; substrate: the support base of the antenna, the IC and the connectors. It contains electrical characteristics that are considered in the design of the antenna.

The inlay can be encapsulated or not. Both the material to be used for the encapsulation and its format will be defined according to the characteristics of the application. Thus, an RFID tag can be provided in many shapes, types and sizes.

Still belonging to the physical layer it is the interrogation zone. The interrogation zone represents the area in which the RFID reader is able to activate and obtain an answer from an RFID tag, one of the crucial points for any implantation.

The RFID software layer, named middleware, goes beyond simply connecting devices. It allows all necessary applications for the RFID system, as filttering, giving hability to manage data and other applications for the user.

3. Attributes for the implementation of the RFID system

The key attributes for implementing the RFID system in a metallic environment and achieving success with the system are: process vision; requirements for implementing the RFID system within the existing process; the necessary resources for the implementation of the RFID system and; finally, the implementation action plan.

3.1 Process view

The entirely process understanding and knowledge of its features in which the RFID system will be implemented is as important as the RFID technology itself. Understanding the challenges of the process and, consequently, make them suitable for the specific application and determine the best RFID system for this process is crucial. Without this essential view of the process, a mistake in the results of the implementation of the RFID system could occur and, consequently, wrong deliveries to the customer.

For example, the RFID system will be used to identify steel coils (steel wire rod), **Figure 2**.

The steel coils are produced in the rolling mill, where they receive a label for their identification of batch and type of material. The identification tag is attached to the coil manually. After weighing and printing the label with the identification, an operator fixes it to the coil by means of a clamp on one of the coil's ties.

Inside the laminator, the rollers are transported by a type C hook until unloading to the transfer carriage. After labeling, the rollers continue, taken by the hook, towards the transfer cart and finally deposited on the bed. The bed is the delivery point of the rolls for logistics. When the rollers are in bed, it is already the responsibility of the logistics department.

Figure 2.
Steel wire rod.

The coils, under the responsibility of logistics, are stored internally in a warehouse and transported through an overhead crane, as outlined in **Figure 3**.

The destination of the stored coils is for sale to the foreign market, so their transportation is done by trucks or train cars.

In short, the RFID system will be used for storage management. The storage management comprises the flow of the coils, their movement, storage, until the order is picking and shipment [2]. However, with the RFID system, picking is a step that will be eliminated, as the system allows the coils to be removed from the storage position and to be directly shipped for transportation, as shown in the flow in **Figure 4**.

3.2 RFID system implementation requirements

After analyzing the process in depth, a thorough approach to analyze the specific need to use the RFID system and is necessary. After this step, design the system architecture solution with the best cost–benefit ratio. Without these basic requirements there may be an unrealistic expectation in the delivery of the RFID project.

Below are some important aspects of the application:

- The RFID system will be implemented to facilitate the logistics processes of receiving, shipping and inventory.
- The RFID system data must be integrated with the logistics management system.
- Logistic processes require the correct storage position within the warehouse.

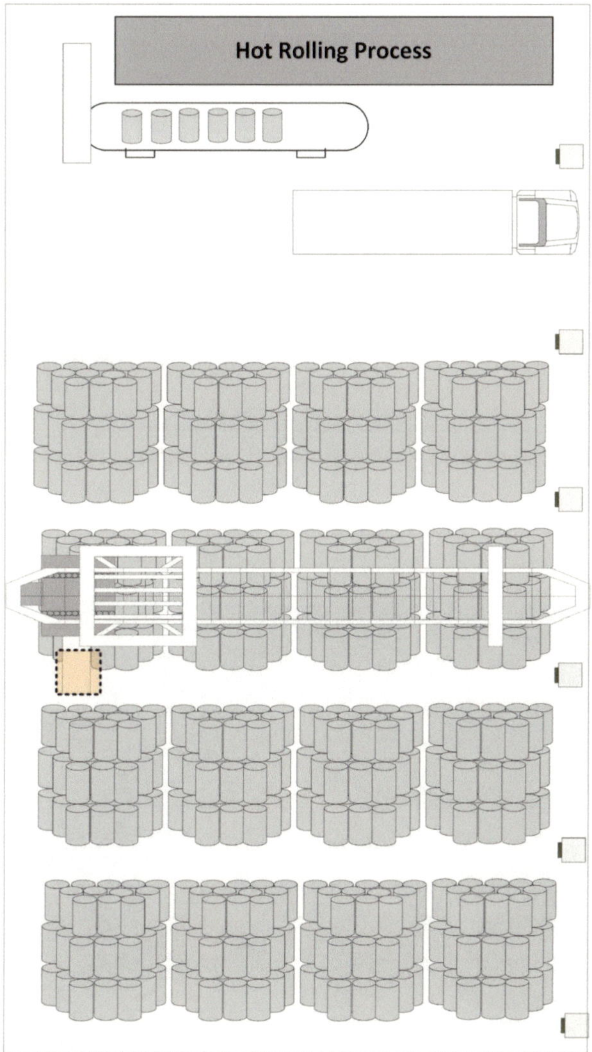

Figure 3.
Storage of steel wire rods.

- The object to be identified: steel coil.
- Steel coils are transported via an electromagnet overhead crane. An operator should never be under a suspended load.
- Steel coils leave the rolling process at high temperatures. The temperature depends on the gauge of the steel. However, coil cooling decreases very fast. **Figure 5** shows the exit temperature of the laminator of a steel coil with 113.5^oC.

3.3 Resources for implementing the RFID system

The resources for the implementation must be mapped so that there is no frustration with the RFID system. The resources are divided into financial and human resources.

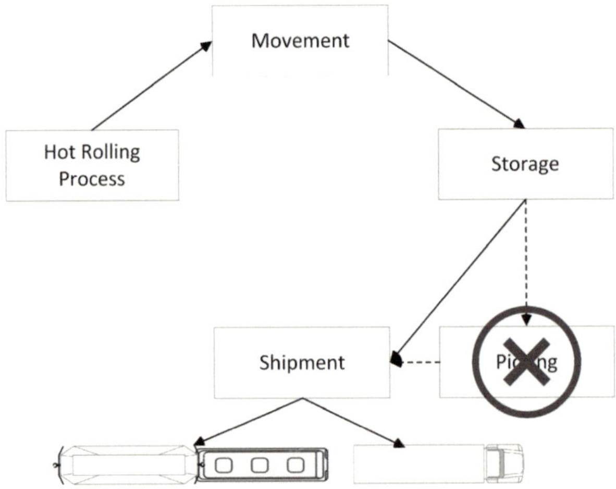

Figure 4.
Storage management flows - with RFID, the picking step is eliminated.

Figure 5.
Coil temperature at the hot rolling process outlet at 113.5°C.

3.3.1 Financial resources

The financial resource is an estimate of the necessary amount of financial resources for the project. The estimated budget is obtained through a business plan based on the architecture of the solution and aligned with the objectives that justify the implementation of the RFID system attending the expected project goals.

The business plan helps to obtain the correct investment decisions and must be focused on defining the problem that the RFID system will solve, considering the objectives and risks that the organization will face when implementing this technology. The definition of objectives at the beginning of the process will ensure that the design of the solution is monitored both during implementation and at the conclusion of the steps and their deliveries; consequently, it will bring subsidies to monitor the evolution of the implementation of the RFID system.

For a steel industry, improving the working conditions and protecting healthcare of their operators are the main objectives for the implementation of the RFID system in coil storage, i.e. the focus is on work safety of their employees. The automatic collection of the coil identification removes the operator from the storage area and considerably reduces the risk of accidents with the coil falling during its transport on the overhead crane.

Business performance can also be scored and should consider: the efficiency of steel coil circulation, inventory management, more effective inventory control, improving operational productivity by reducing management costs, reducing lead times coil loading for road or rail transport.

3.3.2 Human resource

The human resource is the project team needed for the implementation. The RFID system is a complex system and involves hardware and software, so the team must consist of at least:

- 01 radio frequency specialist;
- 01 overhead crane maintenance technician;
- 01 software developer for the development of the RFID Middleware; and,
- 01 project manager

3.4 Action plan for the RFID implementation

An inadequate action plan leads to a false start in the implementation, and as a result, mismatched and uncertain delivery dates. For this reason the action plan must be carefully elaborated and must provide details of all activities involved as who will be in charge and how the piece of necessary information is going to be provided.

The action plan can be divided into four stages: conceptual phase, which describes the initial vision and the basic objectives of the project and the business plan itself; planning phase, which consists of an analytical structure of projects carried out by the project manager, involving responsibility for the work and interrelated tasks; installation phase, which is the stage in which it is possible to identify and purchase all the equipment and software that will be part of the solution. The installation phase will remain until all software and hardware installations in the approval environment are completed, for testing and validation, and later, the turn to the production environment. In this phase, the project manager transfers the project to the operating personnel along with all project documentation, including the service and governance plan.

4. Solution design

In order to analyze the feasibility of implementing the RFID system for identification in steel coils, a proof of concept is required. Proof of concept (POC) is an activity that demonstrates, in the installation environment itself, electromagnetic interference in real conditions and the most appropriate RFID tag for the operation, consequently, the POC identifies the challenges that the implementation of the RFID system will face in this hostile environment for radio frequency.

4.1 The physical layer

4.1.1 RFID tag

The temperature in the steel coils is a variable analyzed in the proof of concept, as it will determine the type of RFID tag for the system.

A coil has three sectors divided into head, middle and tail. The RFID tag must be placed on its tail and on the inside part, named as the cold zone. For this reason, the temperature measurement in the POC must be performed on the steel coil on its tail and inside it at the time of the hot rolling process exit after cooling, according to the points indicated in **Figure 6**, using a FLUKE thermometer.

Table 1 shows the temperatures measured at the exit of the laminator at measuring points 1, 2, 3 and 4, as shown in **Figure 6**, performed at an ambient temperature of 23°*C*.

In addition to the measurement shown in **Table 1**, the coil temperatures must be recorded during the storage process, that is, on the scale, in the storage place after transportation with the overhead crane and after 40 minutes of storage. As an example, the following coil gauges were chosen for measurement: 31.75 mm and 7 mm, as the gauge directly influences the temperature, as well as the chemical

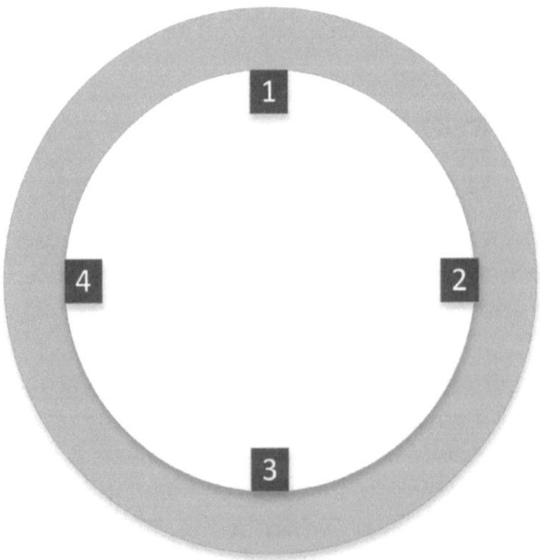

Figure 6.
Temperature measurement points on the steel coil tail after rolling.

Point 1	Point 2	Point 3	Point 4
60.7°*C*	47°*C*	36°*C*	55°*C*
68°*C*	44°*C*	39°*C*	52.1°*C*
55°*C*	41°*C*	35.8°*C*	52.5°*C*
61.7°*C*	50.7°*C*	38.8°*C*	41.3°*C*
60.9°*C*	52.9°*C*	42.5°*C*	50.3°*C*
74.8°*C*	60°*C*	47.3°*C*	58.1°*C*
182.1°*C*	118.8°*C*	117.1°*C*	124.1°*C*
115°*C*	73.2°*C*	97.9°*C*	113.5°*C*
135°*C*	98°*C*	107.6°*C*	119°*C*
158°*C*	102°*C*	108°*C*	98°*C*

Table 1.
Temperature in the coils at the measuring points at the power plant exit.

composition of the steel, among other factors, however, these characteristics were not considered in this work.

After the first 20 laminated pieces, it is suggested to start measuring the temperature of the 04 posterior coils, named X1, X2, X3 and X4, in point 1, to check if there will be a higher temperature in relation to the other measured points, which could occur. **Table 2** shows the coil temperatures of 31.75 mm during the storage process of 04 (four) coils, referring to the location where the coil was during the measurement, the coil and the respective temperature ^oC. **Table 3** shows the coil temperatures of 7.0 mm.

Looking at **Tables 2** and **3**, it can be seen that 100% of the 31,75 mm and 7 mm gauge coils are above the maximum operating temperature desired for the RFID system in all storage locations, as can be seen in the graph of **Figure 7** and this is a challenge for the implementation of the RFID system, as these conditions can affect the reading distance and the reading rates.

The functional and performance requirements of an RFID tag are influenced by the temperature at which it is submitted. An example can be seen in **Table 4** presented in the technical specifications of the Impinj Monza 4 RFID Tag IC [3].

Place	X1	X2	X3	X4
Weight Balance	180°*C*	140°*C*	140°*C*	125°*C*
Overhead crane	136°*C*	118°*C*	116°*C*	106°*C*
min after storage	120°*C*	90°*C*	90°*C*	90°*C*

Table 2.
Coil temperatures of 31.75 mm during storage process.

Place	X5	X6	X7	X8
Weight Balance	199°*C*	202°*C*	220°*C*	238°*C*
Overhead crane	176°*C*	182°*C*	192°*C*	217°*C*
min after storage	100°*C*	140°*C*	123°*C*	134°*C*

Table 3.
Coil temperatures of 7.0 mm during storage process.

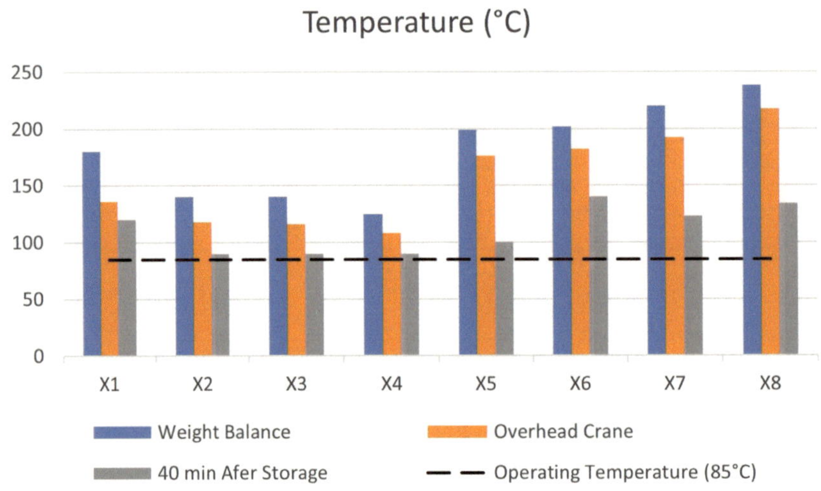

Figure 7.
Temperature of coils X1, X2, X3 and X4 (31.7 mm) and X5, X6, X7 and X8 (7 mm) in the different storage locations and the operating temperature of the RFID system.

Parameter	Minimum	Maximum	Units	Comments
Extended	−40	+85	°C	Default range for all requirements
Operating				functional and performance
Temperature				requirements
Storage	−40	+85/+125	°C	At 125°C
Temperature				data retention is 1 year
Assembly		+150	°C	Applied for one minute
Survival				for one minute
Temperature				

Table 4.
Temperature parameters – Impinj Monza 4 tag Chip.

A solution to the high temperatures of the product is to use RFID tags developed by several suppliers of special tags. **Table 5** presents the specification of some High Temperature UHF RFID Tags developed by some worldwide recognized manufacturers that would the possible to be used in the mentioned RFID application.

However this solution makes the project unfeasible due to the price of these tags, due to the current Brazilian market complex tax system as import duties, taxes, shipping fees, and a few other costs applied when importing goods to Brazil.

Another solution for high temperature is to encapsulate the tag to protect it and prevent damage to the IC or to the connector that connects the antenna to the IC. This solution was designed specifically for the characteristics of the project and developed in the country itself, with no import cost. This was the recommended solution as it enabled the project in the RFID tag requisite.

4.1.2 RFID reader

The RFID reader is a device that reads, writes and processes the data on the tags and sends them to an application. Finally, it is responsible for remotely energizing

Specifications	HID [4]	Omni-ID [5]	Confidex [6]
	High Temperature Label	IQ 800P HT	Heatwave Flag
Operating Frequency (MHz)	865 to 956	860 to 960	865 to 928
Read Range	Up to 8 m	Up to 8 m	Up to 10 m
Application Temperature	+230°*C*	+230°*C*	+230°*C*
Dimensions (mm)	80 × 50 × 0.5	85 × 55 × 0.49	76 × 55 × 0.37
IC type	Monza 4QT	Alien Higgs-3	Alien Higgs-9
Housing Material	Aramid polymer	High temperature synthetic label	Special polymer designed for high temperatures

Table 5.
Physical specifications - UHF RFID tag - High Temperature.

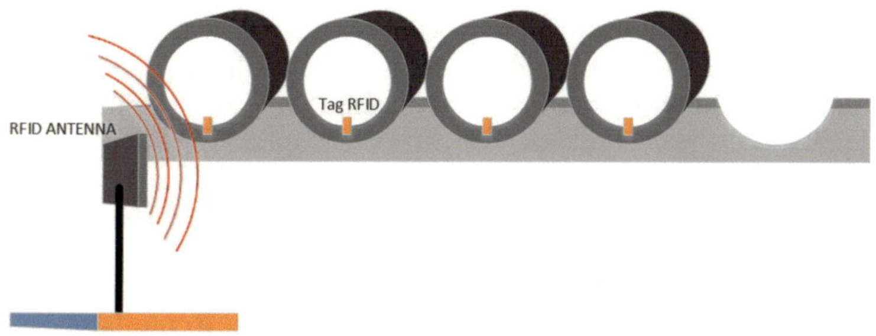

Figure 8.
RFID reader at the output of the laminator.

RFID tags and obtaining the data contained in them. The reader can also perform filtering and data collection functions, and manage equipment input and output.

The reading of the RFID tags on the bed is performed by a fixed reader with an antenna, as shown in **Figure 8**.

The reader must be installed in a metallic distribution board of overlapping NEMA Norm category 4 with ventilation. This protection will prevent the conditions of the environment, high temperatures coming from the laminator, and dust, do not cause the interruption of its operation due to the temperature control. Reinforcing that the RFID reader is an electronic equipment and requires operating temperatures between -20^oC to $+50^oC$ [7].

The automatic receipt of the identification of the coils through the RFID system allows the status of the entry in the logistics sector to be more assertive, with verification of the material of the coil with which it was requested for hot rolling process, in addition to the analysis of the return of non-compliant materials, together with the integration of the WMS system (Warehouse Management System), increasing the quality and speed of the information in order to rationalize and optimize the storage logistics and the management of coil stocks.

Leaving the hot rolling process, in the movement stage, the identification of the coils by the RFID system must be performed by a reader installed on the electromagnet overhead crane itself.

4.2 Environment

4.2.1 Electromagnet overhead crane

A feasibility test of the RFID system must be carried out on the electromagnet overhead crane for the movement of steel coils in the storage area. For this, an RFID antenna must be attached to the overhead crane connected to a fixed RFID reader, in order to the performance tests of reading the RFID tags attached to the steel coil. The number of coils for the test must be according to the number of coils that the bridge's electromagnet car supports. For two coils, the RFID tag reading must be analyzed for these two coils, for four coils, the reading must be analyzed for four coils, and so on. Assuming the reading for two coils, the tests are carried out on two steel coils moved simultaneously with activation of the electromagnet. In this way, the RFID tags of the two coils are read with the system in a static way and with the system in motion.

In order to simulate a real situation, RFID tags must be fixed by the operator himself and in accordance to the process already used for this purpose. Thus, the RFID tags are inside the wire rod as shown in **Figure 9**.

With the system activated, consequently, the electromagnet turned on, **Figure 10**, the analysis of the frequency behavior through a spectrum analyzer occurs, **Figure 11**. Simultaneously, the reading of the RFID tags is obtained. The RFID tag must be read with both in the static and in motion systems.

It is verified in the spectrum analyzer that the magnetic field emitted by the electromagnet of the overhead crane is superimposed on the electromagnetic field of the RFID system. The magnetic field does not interfere with the electromagnetic field of the RFID system.

In **Figure 11** it is also possible to see the hopping channels of the Frequency Hopping Spread-Spectrum (FHSS) technology used in the RFID system according to the rules of the Brazilian regulation established by ANATEL (National Telecommunications Agency). It is important to note that the technology Frequency Hopping (FH) operates with a number of fixed channels stipulated by the country's regulatory agency. In the case of Brazil, 35 hop channels for the RFID system are used in the frequency range of 902 to 907.5 MHz and 915 to 928 MHz.

The FH technology uses these channels to send and receive the signals from the RFID system during a time interval established by the standard and in a sequence of use of each pseudo-random channel, thus reducing the probability of interference from other systems that use the same band of frequency.

Results: there is no interference of the magnetic field in the electromagnetic signal of the RFID system and the reading of the RFID tag is performed successfully, as shown in **Figure 11**. Therefore, the test results are positive in this scenario and the movement of the coils can be carried out by the overhead crane in the inner courtyard. Thus, it completes the storage management cycle by supporting the storage and shipping steps with the RFID system by sending the coil identifications to the logistics management system that the organization uses.

4.3 Software layer

The software layer consists of the RFID Middleware. It is a set of software components that act as a bridge between the components of the RFID system (in

Figure 9.
RFID antenna installed on the overhead crane for the RFID tag reading tests.

this case, RFID readers) and the application (logistics management system) of the organization that will receive the events generated by the RFID system. It needs to provide two main functions: monitoring the health and status of RFID devices, readers and reader antennas; and, manage the infrastructure and data flow specific to the RFID system.

For the system proposed in this chapter, the development of the RFID middleware considers: centralization of the reception of events generated by RFID readers in each equipment in which the RFID system will be installed, overhead crane and laminator bed; filtering and processing messages; interpretation of the message sequence according to the needs of the storage management system; forwarding messages to the storage management system. In addition, a system for monitoring the entire RFID system, including the monitoring of readers and their respective antennas, should also be considered.

The big challenge is to mitigate interference. The interference in the proposed system is considered when there is a reading of the identification of the steel coils that are not the desired ones, called unwanted coils. The interference occurs when there is reflection of the radiofrequency wave signal in the coils stored in the storage area and, consequently, the identification of unwanted coils is read by the RFID system installed on the bridge. This phenomenon is more evident when the overhead crane moves over the high stock, that is, many coils stored with stacking them in several overlapping layers.

Figure 10.
Steel coils with RFID tags lifted by the overhead crane, magnetic field emission in the environment.

Figure 11.
Spectrum analyzer measuring the frequencies radiated in the environment during the RFID reading tests with the electromagnetic overhead crane connected.

4.3.1 Mitigation of interference

The mitigation of interference must be performed at the hardware layer and at the software layer.

In the hardware layer, the interference is mitigated by adjusting the minimum power required to read the identification of the desired steel coils. This is done by adjusting the reader power of each RFID antenna with the static system, overhead crane stopped, loading the coils and positioned on top of the stock.

The software layer, on the other hand, must be responsible to solve the interference that still resulted with the power adjustments. For this, the development of the middleware must contain an established result interpretation logic that eliminates the reading of unwanted coils during the movement of the overhead crane.

The readings that occur in the movement of the overhead crane over the high stock, with the desired coils hoisted by the electromagnet. These readings always contain the identification of the desired coils. However, the reading of the identification of unwanted coils is shown punctually during movement, being eliminated during the journey. The problem is when the route is not long enough for these readings to disappear and not be sent to the management system improperly, and thus, the need for the logic of interpretation of the results. The problem occurs when the path is not long enough for these readings disappear and not be sent to the management system improperly, and thus, there is a need for the logic of interpretation of the results.

The logic of interpretation of the results must be analyzed for each situation of the RFID system, that is, not only when it comes to the RFID solution in overhead crane, but also a logic for each RFID system installed in environments such as the one mentioned in this chapter, a logic for the receiving RFID system, reader installed in the laminator bed and a totally different logic for the RFID system on the overhead crane, which will be true when the system is installed on forklifts, for example.

Therefore, the logic of interpretation of the results must be analyzed case by case when implementing the RFID system in metallic environments and this logic will be responsible for the quality of the delivery of events that are generated by the RFID system and delivered to the storage management system.

5. Conclusions

The implementation of the RFID system is possible in a warehouse that contains steel coils, since the necessary attributes for the successful use of this technology are observed. The attributes are divided into four stages: view of the process; requirements for implementation; the necessary resources; and the mapping of the action plan.

The vision of the process is the first step, and it must occur even before the implementation. It represents one of the most important factors in the whole process, as it presents all the challenges that the RFID system will face and serves as a foundation for the other steps. After analyzing the process, the requirements must be very well mapped for a successful implementation. They must contain the most important aspects for the application of the RFID system, such as: the logistical processes that the RFID system will support; the intrinsic variables of the process, item temperature, storage conditions, for example; type of transport that the item will use; safety conditions for the employees; etc.

Resources are divided into two sectors: financial and human. Without the necessary resources for the implementation of the RFID system, especially in a harsh environment, such as the storage of steel coils, the RFID system, during its implementation, can bring some frustration in the deliveries since there will not be trained people available nor financial support for the project on how to proceed correctly. The last step of the attributes is the action plan. The action plan includes the design of the project, the approvals, both of software and hardware in a test environment and, finally, the delivery of the solution in a production environment.

In case there is high temperature in the item that will use the RFID system, the RFID tag must be carefully studied. There are many market solutions for tags for these items, however, the cost can turn the RFID implementation unfeasible and another effective solution should be studied so that the RFID tag is able to support the variables that involve this type of application.

The RFID Middleware, i.e., the software that has the native function of collecting, filtering and grouping raw data sent from RFID tags and collected by the RFID readers, will also insert some business rules necessary for the implementation. The Middleware is also in charge of dealing with the interference arising from the environment so that a successful RFID implementation may occur. In short, the RFID Middleware plays an extremely important role in the whole context making possible to get success in the RFID implemented system in such a harsh environment when dealing with a technology that uses radio frequency at the frequency of 900 MHz.

Abbreviations

RFID	radio frequency identification
UHF	ultra high frequency
LF	low frequency
HF	high frequency
NFC	near field communication
IC	integrated circuit
USB	universal serial bus
POC	proof of concept
FHSS	frequency hopping spread-spectrum
ANATEL	National Telecommunications Agency

Author details

Renata Rampim[1,2]* and Ivan de Pieri Baladei[2]

1 RF Consulting, Sorocaba, Brazil

2 FAAP, São Paulo, Brazil

*Address all correspondence to: renata@rfconsulting.com.br

References

[1] Rampim R., Internet das coisas sem mistérios: uma nova inteligência para os negócios. 1nd ed. NETPRESS BOOKS; 2016. 127 p. ISBN 978–85–65794-02-2

[2] Rampim R. [et al.], Implementando RFID na cadeia de negócios: tecnologia a serviço da excelência. 3nd ed. EDIPUCRS; 2014. 344 p. ISBN 978–85–397-0350-0

[3] Impinj, Monza 4 Tag Chip Datasheet IPJ-W1510, IPJ-W1512, IPJ-W1513, IPJ-W1535. Impinj, Inc; Version 10.0; 2016.

[4] HID High Temperature Label 2020 https://www.hidglobal.com.br/node/23371 [Accessed: 21 December 2020].

[5] IQ 400P HT 2020 https://omni-id.com/datasheet/781/ [Accessed: 21 December 2020].

[6] Product Datasheet Confidex Heatwave Flag 2020 https://www.confidex.com/wp-content/uploads/Confidex-Heatwave-Flag-Datasheet.pdf [Accessed: 21 December 2020].

[7] Impinj, SpeedwayR Installation and Operations Guide Version 6.4. Impinj, Inc; Version 6.4; 2020.

Chapter 9

RFID Applications in Retail

Narges Kasiri

Abstract

Radio Frequency Identification (RFID) technology is one of the latest product tracking technologies being utilized by retailers. Operations management improvements were among the first recognized applications of this technology earlier in the century. RFID applications in managing retail operations, such as inventory management and control, lead to significant benefits. However, RFID applications are not limited to operations management and go beyond the operations side to offer improvements in other areas in retail such as marketing and managing customers' shopping experiences. In this research, we review the applications of RFID technology in retail since its introduction and how those applications have evolved over the last two decades to help retailers provide omnichannel services to their customers in the current market. We will demonstrate what strategic and tactical factors have helped retailers implement this technology and what factors have slowed down the process of adoption. We will also report on the latest status of the utilization of RFID in the retail sector.

Keywords: RFID, RFID applications, RFID in retail, Retail sector, RFID in retail operations

1. Introduction

Retail stores manage millions of items on a day to day basis to deliver to their customers. Point of Sales (POS) systems with barcodes were among the first technologies used to track products across the supply chain and in stores. Barcodes, as an identification technology, are not utilized at item-level but usually represent a group of products. Retailers need to scan products at pallet level at the point of receiving shipments, in inventories entrance and exit places, and at the POS to keep track of what is coming into and leaving stores [1]. With barcode systems, inventory inaccuracy is created because stores barcode scanning are not always performed at the right time and the right location. This inventory inaccuracy leads to a significant loss at retailers. Retailers needed to explore new ways of tracking their items to lower the inventory inaccuracy and prevent consequent losses. Radio Frequency Identification (RFID) technology appears to be the new technology solution that could improve the inventory record accuracy of stores for various items.

RFID technology applications have been recognized in many areas such as healthcare, finance, manufacturing, and retail. The share of the RFID market in retail is projected to be the largest of all sectors with about 34% by 2026, followed by transportation sector (25%), financial and security services (22%), and other industries such as healthcare and manufacturing at smaller portions [2]. RFID tags can store more information about each item at real time and can have individualized identification for items versus barcodes with a small data storage capacity that can only identify a group of items. RFID readers do not need to be on the line of sight to read RFID tags information which means items can be scanned more frequently and faster at any movement. These capabilities allow little mistakes in tracking records and largely eliminates inventory inaccuracy.

RFID technology's benefits to retailers were identified early at the beginning of the 21st century. However, RFID's applications in retail stores on a large scale took a while to be implemented. This paper reviews utilizing RFID, as an ideal solution to retail operations, since earlier this century and will cover a twenty-year horizon (2001–2020) divided into three equally long periods of 2001–2007, 2008–2014, and 2015–2020.

2. Early introduction and utilization of RFID (2001–2007)

In studies done earlier in this century, RFID was recognized as the next major identification technology to replace barcode systems in the retail industry [3–7]. Barcode systems have been used to track customer purchases, to manage inventory records, and to offer promotion and advertising in retail since 1970 [8]. Barcode tags, however, need to be on the line of readers to be read, a requirement that makes physical inventory counting a labor-intensive task and prevents stores from updating their inventory records frequently and on time. Therefore, with barcode systems, inventory inaccuracy is significant [9]. Inventory inaccuracy refers to the difference between inventory on record and the actual number of items on hand in stores. Inventory inaccuracy is caused by many factors such as transaction errors in the POS system, or shrinkage caused by possible employee/customer theft. Inventory inaccuracy means that stores may not be able to place inventory orders on time, resulting in out-of- stock conditions and consequently losing sales and hurting customer shopping experience. RFID technology, on the other hand, enhances product visibility in store operations and across the supply chain through the ease of reading RFID tag information and updating inventory records on a real-time basis.

Studies have investigated RFID benefits in different areas of retail operations, such as supply chain management, and show how inventory inaccuracy and consequently out-of-stock conditions are improved with the implementation of RFID across the supply chain [10–12]. Enhanced information visibility, provided by RFID in the supply chain, decreases uncertainties and lowers high inventory costs associated with the uncertainties [13, 14].

Many pilot studies during this period investigated and explored the applications of this technology at the pallet level, case level, and item levels in stores [1, 6, 12, 15, 16]. Bottani and Rizzi [1] conducted a case study in 2005 to analyze pallet and case-level implementation of RFID and enhanced visibility generated at the receiving gates and entrance doors from backstore to sales floors. They demonstrated that safety stock and inventory holdings can be significantly reduced and RFID benefits are broad, ranging from labor efficiencies to inventory management improvements. Cost–benefit analyses in this period showed that pallet-level implementations of RFID were more cost effective than case-level implementations.

Metro Group in Germany conducted some case studies in their stores to show item-level RFID applications can improve customers' shopping experiences as well. They introduced some tools provided by RFID technology such as automatic checkout, smart carts that help customers navigate stores and find their items easier and faster, and smart dressing rooms that help customers find their desired apparel items more conveniently [6, 16]. They demonstrated that utilizing these tools significantly enhances customers' shopping experience.

Walmart retail stores in the US were the first retailers that decided to mandate the implementation of RFID at pallet and case level across some of their supply chain in 2005. Walmart also did a pilot study with 24 stores over a period of around 6 months to measure how RFID can improve inventory management. They demonstrated out-of-stock conditions were significantly reduced with the implementation of RFID technology [12].

In Asia, two Singaporean fashion retailers piloted item-level RFID on their apparel stores and reported significant reduction in stocking time from hours to minutes that consequently increased the frequency of counting items with handheld readers and improved inventory accuracy [17].

3. Delays and reflection time on implementing RFID solutions (2008-2014)

Financial crises and the great recession that started in 2008 did not work to the advantage of retailers that were planning to implement RFID applications in their stores. During financial crises, businesses tend to adopt strategies that could help them sustain and survive by spending low and investing less. RFID technology implementation plans were mostly postponed or slowed down during the financial crises. However, this period was the best time to develop some foundations with respect to policies, regulations, and standardization of the technology.

Privacy issues raised by consumer protection agencies and standardization issues across different platforms put forward by case studies and pilot projects led to the development of some regulations and privacy policies by governments, institutions, and businesses. European Commission (EC) took an active role by funding many initiatives across Europe [18]. Initiatives such as Coordinating European efforts for promoting the European RFID value chain (CE RFID) [19] and Building Radio frequency Identification solutions for the Global Environment (BRIDGE) [20], conducted from 2006 to 2009, highlighted that wide implementation of RFID technology needs some regulations, standardizations, and privacy policies in place. For example, the BRIDGE project, coordinated by GS1, helped the industry to develop standardizations such as establishing a common format for the data stored on RFID tags, or the availability of possible frequency bands.

RFID tags can store identifiable consumers' private data, which need to be protected. Therefore, EU members signed an agreement on the Privacy Impact Assessment framework in order to protect consumer privacy [21–23]. This agreement established some rules to be followed in the design of smart chips such as RFID tags to protect the privacy of consumers' data. Consumers should be informed if RFID tags are utilized in stores. In addition, tags must be deactivated at the point of sales at no cost [24]. This framework was later expanded to cover some rules for smart meters as well. In the United States, lawsuits against RFID application patents as well as privacy issues in 2011–2013 were setbacks for large-scale implementations of the technology. The National Institute of Standards and Technology (NIST) in the US has helped to establish some guidelines to help retailers; however, most of the development of policies and standardizations have been initiated by corporations in the US.

In addition to developing policies and standardizations, businesses had more chances to identify and learn broader applications of RFID technology in retail. The focus of most of earlier pilot studies was how this new tracking technology helps manage inventories better in order to avoid out-of-stock conditions. However, the applications of the technology go beyond only inventory management and tracking items throughout the supply chain. As shown in the balanced scorecard developed in [25], RFID benefits extend to marketing and merchandising operations in retail as well (**Figure 1**).

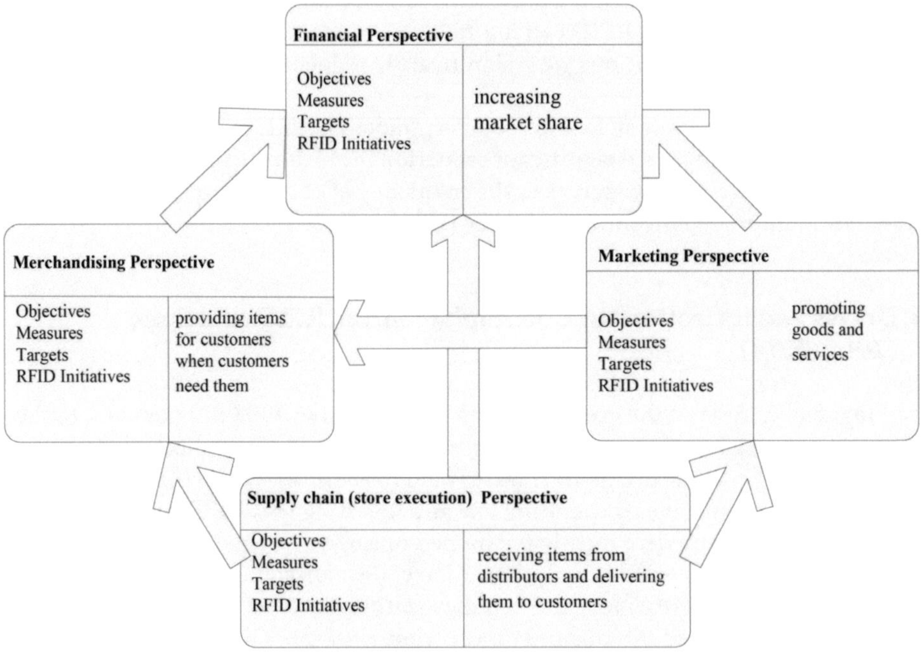

Figure 1.
Balanced scorecard for RFID applications in retail.

In marketing, stores can monitor the behavior of consumers better when customers use tools such as smart carts or smart dressing rooms provided by RFID. Retailers can learn about consumers' preferences and reflect that in the promotion and advertising offered to customers in real time while they shop. The available tools such as smart dressing rooms and smart carts also enhance customer shopping experience. Use of these tools enables customers to find their desired items more conveniently and faster, which eventually leads to higher customer satisfaction and increase in sales.

In merchandising, enhanced visibility on consumer behavior in stores provided by RFID can help retailers identify better assortments of products. In addition, an enhanced visibility means better shelf-replenishment; that is stores can reduce the shelf space since enhanced visibility on shelves allows retailers to replenish them as soon as they become emptied. Less shelf space leads to holding less number of items on shelves at any given time and consequently less inventory and capital held in stores, which allows retailers to invest in carrying more variety for products in stores.

There were also more studies during this period conducting cost–benefit analysis of the implementation of the technology. The fixed cost of implementation includes middleware, fixed antennas, sensors, and readers and the variable cost includes the cost of tags per item. The cost of tags can be added to the cost of each product but then the big question is who has to pay for that cost. Should the cost be transferred to consumers or should that be shared between retailers and manufacturers? The tag cost as the variable cost of utilizing the technology is huge and cost–benefit analysis studies have shown it to be a major barrier to the implementation of the technology during this period. Kasiri and Sharda [26] showed that the cost of tags in item-level implementation of RFID, as the variable cost, is cumbersome. Moreover, the cost can exceed the benefits in some cases depending on the extent to which stores implement RFID applications. The cost barrier was expected to weigh less as the cost of tags became lower over time.

4. Large-scale implementation of RFID (2015-2020)

Surveys of businesses show the implementation of RFID has picked up in this period. A survey of 60 retail executives throughout the United States and Europe showed about 73% of retailers had plans to implement RFID in 2016 [27]. Another survey in 2018 [28], however, showed that 92% of retailers in North America plan on implementing RFID, which is about a 20% increase from the 2016 results.

The cost of implementation has been decreasing over time, as expected, and at the same time retailers have learned how to partially implement the technology. Retailers realized that they do not need to fully implement the technology. In some cases, only tags and hand-held readers are used to add visibility of items in stores without many of the infrastructures such as antennas. Cloud services, on the other hand, have allowed retailers to eliminate some of the middleware cost as well. The leading European fashion stores C & A is one of many retailers that explored lower cost implementations with partial utilization of the technology [29].

RFID platforms can generate big data that are the records of tracking items throughout the supply chain and stores in real time. Businesses need big data and business analytics capabilities to fully utilize technologies [30]. The results of the analysis of such data can help retailers improve their processes such as shelf-replenishment process as well as variety and assortment planning that have been used in the same format for many years. A new timely replenishment process can result in better management of physical space, layouts, and lowering holding costs in stores. Furthermore, a better variety and assortment planning means fulfilling customers' expectations and eliminating unpopular items that releases some capital and allow investment opportunities in other areas in retail.

Competing technologies to RFID have been developed and utilized over time as well. For example, Quick Response (QR) codes give retailers better ability to manage items compared to barcodes, Near-Field Communication (NFC) technology has some capabilities compared to UHF RFID, and most recently Amazon's cameras increase product visibilities in stores for fast checkouts. In addition, retailers have different priorities in investing in new technologies and there is competition for dollars invested in various technologies by retailers. For instance, a retail chain can focus on improving inventory operations, but another retailer may be focused on improving marketing operations and customer shopping experience in stores by developing new apps that can assist customers make decisions during their shopping time in stores. In a different example, Gucci as an Italian luxury brand name does not suffer from inventory inaccuracy issues but their priority is customer shopping experience and they have utilized RFID tags to protect customers against counterfeiting across the supply chain until products reach their customers [31].

Omnichannel retail has been widely available during this period of time. Retailers' customers can shop at any time, in any place, and via any shopping channel. Omnichannel retailing needs accurate inventories and enhanced product visibility more than any other time. Item-level RFID can, therefore, accommodate the needs of omnichannel retail more than other technologies available [32]. In addition, blockchain as the latest technology in retail can provide automatic exchange of product data carried by RFID tags between different partners across the supply chain. The blockchain in retail solutions are currently being studied in a consortium of large retailers such as Nike, Macy's, and Dillard's in an RFID lab at Auburn University [33].

Figure 2.
3-S model for RFID retail adoption.

The 3-S model (substitution, scale, and structure) introduced in [34] discussed and projected different phases of the adoption of RFID earlier when this technology was introduced. Later the 3-S model was adapted by [35] to describe the current stage of retailers' implementation of RFID applications in retail (**Figure 2**). The substitution and scale stages are covering mostly what has been achieved during the three periods discussed in this paper. In the substitution phase, the RFID technology was utilized to replace the applications of barcode systems in tracking products. In the scale phase, the RFID applications are enabling retailers to manage their operations with more accuracy, efficiencies, and at a higher speed and scale. The structure phase, that is re-engineering processes and completely overhauling retail operations, is still underway. The RFID technology will enable retailers to accomplish things they could not imagine before and allows retailers to tap into completely new domains and applications.

5. Conclusion

Retailers have different needs based on the way they operate in stores. Some retailers must manage large inventories in stores. An enhanced visibility on their products help them improve inventory accuracies and avoid out-of-stock and increase their efficiencies. On the other hand, some retailers have small backstore inventories and every item they receive is put directly on their shelves and available to their customers. Inventory management is not their priority, but they need to focus more on customer shopping experience. Therefore, the enhanced visibility of items in stores is expected to promote retailers' marketing operations. Depending on the way retailers operate and what their priorities are, retailers have to plan on implementing appropriate applications of RFID technology.

As discussed in this paper, RFID has been utilized broadly with various applications. As a revolutionary technology, RFID's implementation can go beyond improving the current processes in retail operations. The current processes can innovatively change to debut completely new applications that are only possible with the enhanced visibility of items in real time. The ensuing big data that is derived from the visibility provided by RFID tags can be analyzed, leading to innovations in retail operations.

Implementing item-level RFID needs to be part of omnichannel strategy in the retail sector. With the wide-spread usage of online retail services such as Amazon, the competition in retail is tougher than ever before. In omnichannel services, retailers need to grant their customers easy access via different channels and make their products available in a variety of delivery services. The accessibility and fast delivery will not be possible with the level of visibility provided by barcode systems. Utilization of technology in retail is evolving quickly and RFID technology is the one that can definitely help retailers win in this overhaul.

DOI: http://dx.doi.org/10.5772/intechopen.95787

Author details

Narges Kasiri
School of Business, Ithaca College, Ithaca, NY, USA

*Address all correspondence to: nkasiri@ithaca.edu

References

[1] Bottani, E., Rizzi, A. Economical assessment of the impact of RFID technology and EPC system on the Fast Moving Consumer Goods supply chain, International Journal of Production Economics, 2008, 112(2), 548-569.

[2] RFID Forecasts, Players, and Opportunities 2016-2026. IDTechEx Annual RFID Report, 2016.

[3] Kambil A. and Brooks J. D., Auto-ID across the value chain: From dramatic potential to greater efficiency and profit. Technical report, Auto-ID center, 2002.

[4] Chappell G. , Durdan D., Gilbert G., Ginburg L., Smith J., and Tobolski J. Auto-ID on delivery: The value of Auto-ID technology in the retail supply chain. Technical report, Auto-ID center, 2003.

[5] Loebbecke, C. Modernizing Retailing Worldwide at the Point of Sale. Management Information Systems Quarterly Executive (MISQE), 2004, 3(4), 177-187.

[6] Loebbecke, C., Wolfram, G. Taking Content Integration to the POS: Enhancing Shopping Convenience in Metro's 'Future Store'. Society of Information Management (SIM) Paper Award Competition, 2004.

[7] Wang, Y. M., Wang, Y. S., & Yang, Y. F. Understanding the Determinants of RFID Adoption in the Manufacturing Industry. Technological Forecasting and Social Change, 2010, 77(5), 803-815.

[8] Fraza, V. Ending inventory errors in 60 days. Modern Materials Handling, 2000, 55(3), A11.

[9] DeHoratius, N., and Raman, A. Inventory record inaccuracy: An empirical analysis. Management Science, 2008, 54 (4), 627-641.

[10] Atali, A., Lee, H., and ÖlpOzer, O. If the inventory manager knew: Value of RFID under imperfect inventory information. White paper, Stanford University, 2005.

[11] Heese, H. S. Inventory record inaccuracy, double marginalization and RFID adoption. Production and Operations Management,2007, 16(5), 542-553.

[12] Hardgrave, B. C., Langford, S., Waller, M., and Miller, R. Measuring the impact of RFID on out of stocks at Wal-Mart. MISQE, 2008, 7(4), 181-192.

[13] Delen, D., Hardgrave, B. C., and Sharda, R. RFID for better supply-chain management through enhanced information visibility, Productions and Operations Management Journal, 2007, 16(5), 316-324.

[14] Zhou, W. RFID and item-level information visibility. European Journal of Operational Research, 2009, 198(1), 252-258.

[15] Sharma, A., Thomas, D., & Konsynski, B. R. Strategic and Institutional Perspectives in the Evaluation, Adoption and Early Integration of Radio Frequency Identification (RFID): An Empirical Investigation of Current and Potential Adopters. Proceedings of the 41st Annual Hawaii International Conference on System Sciences, Waikoloa, Big Island of Hawaii, HI, 2008.

[16] Krafft, M., Mantrala, M. K. Retailing in the 21st Century: Current and future Trends. 2006, New York:Springer.

[17] Swedberg C., Two Singapore Fashion Retailers Use RFID to Track Inventory, 2008, RFID Journal,

https://www.rfidjournal.com/two-singapore-fashion-retailers-use-rfid-to-track-inventory, accessed December 2020.

[18] Wolfram, G., Birgit G., and Peter G., eds. The RFID Roadmap: The Next Steps for Europe. 2008, Springer Science & Business Media.

[19] CORDIS. CE RFID. Coordinating European efforts for promoting the European RFID value chain, 2008, http://cordis.europa.eu/project/rcn/80154_en.html (2008, accessed July 2020).

[20] CORDIS. BRIDGE. Building Radio frequency Identification solutions for the Global Environment. 2009, http://cordis.europa.eu/project/rcn/93609_en.html (2009, accessed 18 July 2020). John Edwards Europe Is Rolling Out RFID. http://www.rfidjournal.com/articles/view?10000/2 (2012, accessed July 2020).

[21] European Commission 2. Small chips with big potential: New EU recommendations make sure 21st century bar codes respect privacy. http://europa.eu/rapid/press-release_IP-09-740_en.htm?locale=en (2009, accessed November 2020).

[22] European Commission 3. Privacy and Data Protection Impact Assessment Framework for RFID Applications. http://cordis.europa.eu/fp7/ict/enet/documents/rfid-pia-framework-final.pdf. (2011, accessed 18 July 2020).

[23] European Commission 4. Digital privacy: EU-wide logo and "data protection impact assessments" aim to boost the use of RFID systems. http://europa.eu/rapid/press-release_IP-14-889_en.htm. (2014, accessed November 2020).

[24] European Commission 1. Commission launches public consultation on radio frequency ID tags. http://europa.eu/rapid/press-release_IP-06-289_en.htm?locale=en. (2006, accessed November 2020).

[25] Kasiri, N., Sharda, R., Hardgrave, B. C. A Balanced Scorecard for Item-Level RFID in the Retail Sector: A Delphi Study, European Journal of Information Systems, 2012, 21(3), 255-267.

[26] Kasiri, N., & Sharda, R. Real options and system dynamics for information technology investment decisions: Application to RFID adoption in retail. ACM Transactions on Management Information Systems (TMIS), 2013, 4(3), 11.

[27] Unger, R. & Sain, J. Kurt Salmon RFID in retail study, 2016, https://easyscan.dk/wp-content/uploads/2018/03/rfid-retail_study_-kurt-salmon.pdf, accessed 6 August 2020.

[28] Sain J., Wong A. Transforming modern retail: Findings of the 2018 RFID in retail study, 2018, https://www.accenture.com/us-en/insights/strategy/rfid-retail, accessed December 2020.

[29] Swedberg C., C&A Rolls Out RFID to All Its French Stores, RFID Journal, 2016, https://www.rfidjournal.com/ca-rolls-out-rfid-to-all-of-its-french-stores, accessed November 2020.

[30] Lee, H. L. Big data and the innovation cycle. Production and Operations Management, 2018, 27(9), 1642-1646.

[31] Rae H., Stop Thief! These New RFID Tags Could Help Luxury Clothing Brands Guard Against Theft, 2016, Bloomburg, https://www.forbes.com/sites/haniyarae/2016/08/11/rfid-brands-theft/?sh=27298d3a18de, accessed December 2020.

[32] Kumar, A., & Ting, P. RFID & Analytics Driving Agility in Apparel

Supply Chain, 2019, https://dspace.mit.edu/handle/1721.1/121286, accessed December 2020.

[33] Graen M., CHIP: Chain Integration Project, 2020, Auburn RFID Lab White Paper.

[34] Lee, H.L. Peering through a glass darkly, International Commerce Review, 2007, 7(1), 61-68.

[35] Kasiri, N., Erickson, S., Wolfram, J., A Comparative Study of RFID Adoption Patterns in Retail. 2020, Under Review.